Life and Its Future

Josephine C. Adams · Jürgen Engel

Life and Its Future

 Springer

Josephine C. Adams
School of Biochemistry
University of Bristol
Bristol, UK

Jürgen Engel
Biophysikalische Chemie
Biozentrum of the University of Basel
Basel, Switzerland

ISBN 978-3-030-59074-1   ISBN 978-3-030-59075-8   (eBook)
https://doi.org/10.1007/978-3-030-59075-8

This Springer imprint is published by the registered company Springer Nature Switzerland AG
The registered company address is: Gewerbestrasse 11, 6330 Cham, Switzerland

Preface

The authors are biologists who work on the functions of the extracellular matrix, also called connective tissue. Josephine Adams is Professor of Cell Biology at the University of Bristol, UK, and Jürgen Engel is Professor Emeritus of Biophysical Chemistry at the University of Basel. In previous years we collaborated in research and published some of our results together. The current book does not deal with our research on extracellular matrix, but with a much broader theme, namely the origin and evolution of life, dangers to life on Earth that could threaten or even terminate it, the possibility that life may exist on other planets, and future prospects for humans to achieve artificial life.

Why have we written this book? First of all, as bio-scientists, we are deeply engaged with the living world, the biosphere, through our education and areas of research. We like the complexity, diversity, and beauty of life on earth. As humans, we share this feeling with many other people on our planet. Nature provides oxygen, water, food, and the living environment and is essential for human beings in many other ways. The possibility of destruction of life by nuclear warfare has been present throughout our working lives yet has been managed by international treaties. However, the flood of data that the biological world is widely endangered by the general activities of humans, and the pace of this change, are shocking. There is now strong evidence for human-made climate change with global heating that has led to increased public concerns and demonstrations by young people around the world. During the time that we wrote this book, more data has emerged on loss of species and habitats and there have been major losses of the Amazon rainforest through fires followed by its conversion to farmland. Currently, the human way of life is challenged by the Covid-19 pandemic and there are indications that the modern lifestyle and the size of the human population will make pandemics more likely in the future.

In parallel, visions of artificial life-forms composed of inorganic compounds, which could be connected to computer hardware and governed by artificial intelligence, have emerged. Max Tegmark, an expert in this field, shocked us by his statement "If future superhuman artificial intelligence becomes the biggest event in human history, then how can we be sure that it doesn´t become the last?"

We decided to collect scattered information into a short book that may help us and others obtain an overview of these complex and emotive issues. The book begins with a review of current scientific models for the origin of life, the evolution of living organisms and how scientific progress has made possible the technologies of genetic engineering that enable artificial evolution. These chapters summarize current knowledge as a necessary basis for the discussion of attempts to create artificial life. In the second part, we turn to risks to life, either through natural events, or from events of human origin; the latter already contributed historically to the destruction of species (for example, the dodo and passenger pigeon), and now threaten to change the global climate, ecosystems, and the diversity of life as we know it on planet Earth. The discussions include personal reflections and reminiscences of life events from Jürgen Engel. We continue on to the question of artificial intelligence, and end with discussion of whether there may be life forms on other planets and whether it could become feasible for humans to live on other planets. In 1984, the eminent biologist Edmund O. Wilson coined the term "Biophilia" in his book of the same name to capture the idea that humans have an innate tendency to seek connections with other forms of life in the natural world. We hope that the book will help thinking about these issues and also stimulate goodwill and biophilia. We thank the publisher Springer and welcome readers to send corrections of fact.

Bristol, UK   Josephine C. Adams
Basel, Switzerland   Jürgen Engel

Acknowledgments We are grateful to pastor Doris Engel Amara for helping to write Chap. 2, Prof. Ulrich Quast for input on Chap. 6, Dr. Richard Adams for input and comments on Chaps. 6, 8 and 9, and Prof. Hans-Peter Bachinger and Frank Wals for commenting on versions of the text.

Contents

Chapter 1
Introduction

1.1   What Is Life?

We are part of the biocosmos, a web of many millions of living species on planet Earth that range from tiny bacteria and plankton to elephants, whales and giant redwood trees. The origin of life on Earth is perhaps the most fascinating, unsolved problem in biological science. How this took place billions of years ago is not known but a number of plausible models have been proposed and will be discussed in this book. Living systems changed our planet: some rocks originate from the biomineralization of corals and other animals and the oxygen level in the early atmosphere was strongly increased by the photosynthetic activities of certain bacteria, and later by algae and plants.

Before going further, we should address the question: What is life? In ancient times, our ancestors believed that only gods with supernatural powers could have created the stars in the sky and life on Earth. This belief is maintained in different religions, and we discuss some creation myths in Chap. 2.

With regard to a scientific definition of life, although knowledge of chemistry, physics and the biological world has progressed enormously in the last century, it is remarkable that lively debate continues over a formal definition of life. A standard dictionary definition of life would state "the condition that distinguishes animals and plants from inorganic matter, including the capacity for growth, reproduction, functional activity, and continual change preceding death". However, this type of definition does not embrace entities such as viruses (which can only grow and reproduce within a host) and is questionable for organisms (e.g., cnidarians such as *Hydra*) that may be viewed as immortal due to their capacity to undergo asexual reproduction.

Other definitions have emphasised chemical attributes and the process of evolution. The definition by the USA National Aeronautics and Space Administration (NASA) states: "Life is a self-sustained chemical system capable of undergoing Darwinian evolution (Joyce et al. 1994)". The inclusion of Darwinian evolution avoids inclusion of non-living chemical systems such as crystals that can "reproduce". However, as discussed by Cleland and Chyba (2002), some problems remain

J. C. Adams and J. Engel, *Life and Its Future*,
https://doi.org/10.1007/978-3-030-59075-8_1

with this definition. For example, inter-species hybrids that are sterile are excluded. We also do not know if extra-terrestrial life (if ever discovered) would undergo Darwinian evolution. More recently, Higgs (2017) proposed a refinement of the NASA definition to "Life is a self-sustained chemical system capable of undergoing biological evolution". This version draw a distinction between the chemical evolution of self-replicating molecules that is determined by their physicochemical properties and the biological evolution of living organisms that brings vast possibilities to evolve new genes and adapt to environmental changes. Another recent definition also starts from the NASA definition but incorporates an emphasis on thermodynamics "Life is a far from equilibrium, self-maintaining chemical system capable of processing, transforming and accumulating information acquired from the environment" (Vitas and Dobovisek 2019). We will discuss ideas and models on how life originated in Chaps. 3 and 4.

As far as is known at present, life on Earth is a singular phenomenon. To date, there is no evidence for the existence of life on other planets. The mechanisms of life's origin and its evolution to its present state of complexity are of high interest for many reasons. A deep understanding of life on Earth is of value in its own right. This knowledge may help us to conserve ecosystems in the face of current threats caused by the large modifications of our planet that have been made by humans. Models for the origin of life on Earth may assist more effective searches for evidence of life elsewhere in the Universe. The growing fields of synthetic biology and artificial intelligence consider the possibilities of generating "artificial life" which could include super-intelligent machinery, robots or artificial intelligence networks that could potentially go out of human control. The latter possibilities raise discussions about the nature of consciousness and responsibility and many ethical issues for the future.

1.2 Evolution of Life on Earth

Life on Earth developed over a very long period. This is proven by the evidence of fossils of different organisms encased in geological rock layers of different ages. The oldest known fossils include mats of bacteria-like cells (stromolites) detected in rocks dated to be 3.4 billion years old in western Australia (Schopf et al. 2018). This places the origin of life at an earlier time: after 5.5 billion years ago, when our planet was formed, and before 3.4 billion years ago when the already complex bacterial cell communities existed. Based on current knowledge of the types of molecules from which cells are formed, the origin of life is thought to have involved interactions of molecules that gained the capacity to be self-perpetuating. There are no fossils of these tiny precursors of living cells, but plausible models for the chemistry and physics of their formation have been developed. We discuss several prominent models in Chap. 3.

The evolution of life from molecular precursors to the three domains of life known today (bacteria, archaea and eukaryotic cells) would have been guided by the changing conditions on planet Earth as well as by changes in the genetic content of organisms. In Chap. 4 we briefly outline the theory of biological evolution that was initiated by Charles Darwin in the nineteenth century and the significance of deoxyribonucleic acid (DNA) as the molecule of inheritance. In Chap. 5 we discuss modern molecular biology methods that allow the genetic content of a cell or multi-cellular organism to be manipulated artificially. The plasticity of biological organisms (i.e., their ability to adapt to changing environmental conditions) has not only allowed the great diversification of life on Earth that we enjoy and benefit from today, but has also allowed life to continue, in one form or another, after major catastrophic natural events on Earth. We discuss some examples of these natural disasters and their consequences in Chap. 6. Along with pandemics, these include prior mass extinctions for which there is knowledge from the fossil record, geochronology (scientific methods to determine the age and history of rocks and sediments) and paleo-climatology.

1.3 Influence of Humans

Natural selection and evolutionary changes take place over many generations. The first bipedal ancestors of humans emerged in Africa over 4 million years ago and started to migrate out of Africa around 2 million years ago. This is a tiny fraction of the time that life has existed on Earth (about 0.05%), yet the effects of humans on the planet and on other life forms have been profound. Human have a high ability for using tools and new inventions, and for passing on knowledge through language and social practices. In the last several thousand years of human existence, large cultures and urban centres were formed. Forest clearance and tilled fields, villages, cities, and constructions such as the Pyramids, Stonehenge or Mayan temples are among the signs of purposeful human-made changes to the environment. Over the last two hundred years, since the start of the first Industrial Revolution, the activities of humans have brought about increasingly rapid, large-scale changes to the natural environment. Over this time, inventions such as canals, railroads and trains, cars, aircraft, electric power grids, nuclear power stations, telecommunications networks, factory farming and many others have continued to emerge and resulted in profound changes to the natural landscape. Deep insights in different fields of science into the history of our planet and life on Earth have also been made. Now, in the twenty-first century of our human calendar, only few remote areas of our planet remain untouched from human influence. With the discovery of microplastic particles (fragments of plastic items that are <5 mm long) polluting polar waters, snow, animals and the deep seas (Waller et al. 2017; Bergmann et al. 2019; Choy et al. 2019), it is in fact questionable whether any parts of the Earth remain "untouched". As witnessed by current evidence for decreasing biodiversity and increasing species loss around the world, the mechanisms of Darwinian evolution cannot cope with the very rapid and diverse environmental changes that are taking place through human activities.

In 1962, the publication of Rachel Carson's book "Silent Spring" drew attention to the damage to the environment that had been caused by widespread use of pesticides. This led to bans on the use of dichlorodiphenyltrichloroethane (DDT) as a pesticide.

Awareness of the problems of environmental changes became more widespread in the 1970s. In 1972, a famous report "Limits to growth", was presented by the Club of Rome, which made alarming predictions based on the development of industry, the growth of the human population, the lack of food supply and the limits of non-renewable natural resources. The first World Climate Conference was held in Geneva in 1979 and set up the goal of achieving agreed international actions against these risks. It is now appreciated that risks also arise from the use of biological techniques by which the hereditable genetic makeup of organisms can be manipulated, and we touch on these in Chap. 5. A darker side of scientific progress in the twentieth century has been the development of very dangerous weapons of war, with the potential to poison or pollute large areas, or cause widespread destruction through nuclear radiation. The risks of these human inventions and activities are the subjects of Chap. 7. Beyond these unquestionable risks, a general risk to life on Earth has arisen because of changes to the planet's climate driven by global heating due to the large-scale emission of greenhouse gases from human activities. This problem has assumed a massive urgency over the last century. We discuss in Chap. 8 the current evidence that human activities have led to the rapid changes in global climate and provide examples of the devastating effects on other species and their ecosystems.

Many ways to counteract the risks of environmental catastrophe have been proposed. Two types of proposals may be distinguished. The first type deals with measures to moderate the human life style, or to restrict use of fossil fuel resources to a state that would be sustainable. The second type of proposal is based on the value of new inventions and technology to solve problems. For example, some researchers propose artificial intelligence as the basis for generally improving human life on Earth. Others suggest that technology to allow emigration to Mars could be developed. The writings of the astrophysicist Sir Martin Rees (2017, 2018) and other discuss how super-intelligence might help humans to explore the universe. These visions also include the concept of followers of humans, which would be composed in part of inorganic compounds and live linked to artificial intelligence systems. In Chap. 9 we discuss artificial intelligence and the feasibility of designing new inorganic life. In Chap. 10 we discuss the long-standing question of whether there are other life forms in the Universe and the feasibility for humans or other organisms from Earth to colonise another planet.

1.4 Looking to the Future

Overall, a broad aim of this book is to discuss life and its future in view of the many risks that threaten it. Since 1945, the most overtly dangerous risk to life has been from nuclear weapons. With the progress made in artificial intelligence, along with the benefits, there are many new possible scenarios in which the use of super-intelligent robots could "go out of control". The possible consequences pose risks we cannot yet fully understand; for example, the risks of using AI robots as artificial soldiers. Many responsible scientists and philosophers have begun to express alarming views about the future of life. For example, Sir Martin Rees' book raised the question "Will civilisation survive the twenty-first century?" (Rees 2003).

Since 2018, public demonstrations for governmental initiatives to swiftly reduce greenhouse gas emissions and counteract climate change have become increasingly prominent around the world, especially during 2019. However, in 2020, the appearance of the covid-19 pandemic has also made us all aware of the massive risks to human life that are posed by new infectious agents. In Chap. 11 we discuss recent global risk assessments and views on the most pressing risks to life, both to human life and in general, over the next decade.

References

Bergmann M, Mützel S, Primpke S, Tekman MB, Trachsel J, Gerdts G (2019) White and wonderful? Microplastics prevail in snow from the Alps to the Arctic. Sci Adv 5:eaax1157 https://doi.org/10.1126/sciadv.aax1157

Carson R (1962) Silent spring. Houghton Mifflin Company, Boston, USA

Choy CA, Robison BH, Gagne TO, Erwin B, Firl E, Halden RU, Hamilton JA, Katija K, Lisin SE, Rolsky C, Van Houtan KS (2019) The vertical distribution and biological transport of marine microplastics across the epipelagic and mesopelagic water column. Sci Rep 9:7843. https://doi.org/10.1038/s41598-019-44117-2

Cleland CE, Chyba CF (2002) Defining 'life'. Orig Life Evol Biosph 32(4):387–93. https://doi.org/10.1023/a:1020503324273

Higgs PG (2017) Chemical evolution and the evolutionary definition of life. J Mol Evol 84:225–235. https://doi.org/10.1007/s00239-017-9799-3

Joyce GF, Deamer DW, Fleischaker G (1994) Origins of life: the central concepts. Jones and Bartlett, Boston

Rees M (2003) Our final hour. Basic Books

Rees M (2017) Unsere Nachfahren werden Maschinen sein, Neue Zürcher Zeitung October 2017

Rees M (2018) On the future: prospects for humanity. Princeton University Press

Schopf JW, Kitajima K, Spicuzza MJ, Kudryavtsev AB, Valley JW (2018) SIMS analyses of the oldest known assemblage of microfossils document their taxon-correlated carbon isotope compositions. PNAS 115:53–58. https://doi.org/10.1073/pnas.1718063115

The Club of Rome (1972) The limits to growth. https://www.clubofrome.org/report/the-limits-to-growth/

Vitas M, Dobovisek A (2019) Towards a general definition of life. Orig Life Evol Biosph 49:77–88

Waller CL, Griffiths HJ, Waluda CM, Thorpe SE, Loaiza I, Moreno B, Pacherres CO, Hughes KA (2017) Microplastics in the Antarctic marine system: an emerging area of research. Sci Total Environ. 598:220–227. https://doi.org/10.1016/j.scitotenv.2017.03.283

Chapter 2
Early Ideas About the Origin of Life

It is clear that humans have been keen observers of nature since early pre-history. The paintings which early humans have left in the prehistoric Chauvet-Pont-d´Arc cave in the Ardèche department of France show that observations and reflections were very highly developed about 40,000 years ago. The paintings are highly impressive pieces of art showing a deep understanding of animals. In addition, abstract paintings with magical and ritual aspects are seen in this cave. The world's oldest continuous cultural tradition is that of the Aboriginal Australians, who entered Australia over 50,000 years ago (Tobler et al. 2017). Aboriginal Australians have a very complex culture and belief system based on oneness with nature, that is communicated through the generations by rock art, sand painting, wood carving, ritual body painting and other art forms as well as by oral tradition. In this world-view, all forms of life are seen as inter-linked and ways of conduct come from the Dreamtime, when the world and all living beings were created by eternal, mythical beings.

Written documents relating to observations of nature are much younger, a few thousand years old at most. Very interesting are the myths of creation from different parts of the world. Almost all of these myths were originally communicated orally in earlier times. A nice summary of different creation myths has been published by Steinwede and Förster (2004).

Creation myths document the great interest of early humans to explore the properties of the natural world that they were facing in their lives. Speaking in modern language, many of the wonders they observed fall into the areas of astronomy, geology and physics. Examples are the rising and setting of the sun and the moon, the movements of stars, the tides, and the shape of the Earth. Other profound questions relate to biology and the origin of life, the enormous diversity of life on Earth, the changing seasons, the mystery of death, and the capacity of humans for good and evil.

The myth of the Ojibwa Indians of the northern USA and Canada (slightly shortened from Steinwede and Förster (2004)) states:

J. C. Adams and J. Engel, *Life and Its Future*,
https://doi.org/10.1007/978-3-030-59075-8_2

> *Who gave me the breath of life, my shape of flesh?*
> *Who gave me the beating of my heart and the seeing of my eyes*
> *Who?*
> *Who gave the rose the charm of her form,*
> *the beauty of her colours?*
> *Who gave the pine tree the secret of growth,*
> *the power to heal?*
> *Who gave the bear the sense of time,*
> *a place to stay during the winter?*
> *Who gave the sun the light to shine, her path to find?*
> *Who gave the earth its green richness, the cycles of being?*

In this myth, questions are asked and no explanations are given.

The famous Sanskrit myth "Rigveda", that was written between 1500 and 1250 years before Christ, introduces the concept of Gods as creators but only in a rather reserved way (this excerpt from Steinwede and Förster, (2004)):

> *No death was there and no life,*
> *no sun, no moon and no stars*
> *but then started the being, there was breathing*
> *darkness was in the world, the universe a great chaos,*
> *then started life, a seed, a beam, born by the power of fire.*
> *who is certain about how creation originated?*
> *at the side of creation were Gods.*
> *where did they come from?*
> *who knows, how all this happened*
> *and whether it happened by intention.*
> *a highest God in heaven knows it or does no know it?*

The Christian Old Testament draws a very clear picture of creation (paraphrased in own words):

> *At the first day the earth and the universe were created,*
> *plants and animals followed,*
> *the human species was created separately,*
> *females originate from the rib of men.*
> *most of this creation happened in seven days*

Creation myths are beautiful pieces of poetry. They tell us about the attempts of our early ancestors to explain the world into which they were born. If living at the time of the paintings in the Chauvet-Pont-d´Arc cave, people were immersed by the beauty and risks of nature and the struggle for their own survival. They considered that only wise and powerful Gods could have been able to create the wonders of the world.

Many myths also deal with the cruel aspects of nature and introduce Gods for good and for evil. An example is the Persian myth of Zarathrustra (= camel guide) from Steinwede and Förster (2004), which is dated to 1000–500 years before Christ
:

The bisexual Maingod Zurvan gave birth to two sons.
Ahura Mazda for truth and light and Ahriman for all evils.
Zurvan had to accept both sons, because good and evil will always act together.

The creation myths are the products of human imagination. They are at variance with modern science. A clear misconception is the assumption of relatively rapid time scales for the formation of the universe and the creation of life, and the belief that both events occurred close together in time. Many of the creation myths became incorporated into more recent religions and gained influences that continue to the present day. A belief that some of the myths are literally true is termed creationism. Religions have completely different aims to science and in the rest of this book we will focus on scientific evidence of how life began on Earth, current threats to life on Earth and the possibility of life on other planets.

References

Steinwede D, Förster D (2004) Die Schöpfungsmythen der Menschheit. Patmos Verlag
Tobler R, Rohrlach A, Soubrier J, Bover P, Llamas B, Tuke J, Bean N, Abdullah-Highfold A, Agius S, O'Donoghue A, O'Loughlin I, Sutton P, Zilio F, Walshe K, Williams AN, Turney CSM, Williams M, Richards SM, Mitchell RJ, Kowal E, Stephen JR, Williams L, Haak W, Cooper A (2017) Aboriginal mitogenomes reveal 50,000 years of regionalism in Australia. Nature 544:180–184. https://doi.org/10.1038/nature21416

Chapter 3
The Scientific View of the Origin of Life

3.1 The Early Earth

Our current knowledge about the origin of life is based on several different fields of science. Astronomy has contributed the knowledge to estimate the time at which our planet was formed and what chemical constituents were present at that time. Geology has provided information on the planet Earth. Chemists and biochemists have made predictions about the prebiotic chemistry on which the origin of life was based and palaeontology has contributed the study of fossilised life forms. In addition, laboratory experiments have been devised to mimic conditions on the early Earth or to study proposed evolutionary processes under controlled conditions. As far as we know, life has originated only on planet Earth.

The Earth was formed by accretion of particles from dust clouds of the Sun sometime between 5 and 4.5 billion years ago. The early Earth was a very dynamic structure with molten surface rocks (magma), extensive volcanic activity and rapid variations in the chemical composition of its atmosphere. The atmosphere of the young Earth was probably too thin to filter the heavy flux of cosmic radiation. Therefore natural UV irradiation from sunlight and radioactivity reaching the surface of the Earth would have been much higher than today and would have had large influences on the chemical composition of Earth. The Earth was also exposed to heavy asteroid bombardments (an asteroid is a rocky body orbiting the sun that does not correspond to a planet), which led to dramatic but poorly understood changes in the planet, its atmosphere and geology. From ongoing space missions to collect surface material from asteroids—for example the successful return of material from asteroid Ryugu by the Hayabusa-2 space probe (https://www.hayabusa2.jaxa.jp/en/)—we will learn more about the early Solar System and Earth. Earth's water may have come from the materials from which the Earth was first formed, or could have also been delivered by the asteroid bombardments (Peslier 2020). As an atmosphere was formed, possibly from gases released by volcanoes, sunlight at the Earth's surface was reduced, leading to reduced surface temperature, cooling and solidification of the surface rocks and

J. C. Adams and J. Engel, *Life and Its Future*,
https://doi.org/10.1007/978-3-030-59075-8_3

condensation of water into the oceans. Once these stages had begun, the Earth became a possible place for life to emerge.

The age of the Earth is another topic of active research. Scientific research into the chemical properties of rocks has built up detailed information about the timeline of rock formation back to about 4 billion years ago. There is no definite record for earlier geological stages, as rocks that existed before 4 billion years ago have tended to be obliterated or modified by later events. For example, the Earth's Moon was formed from the impact of an asteroid on the early Earth about 4.6 billion years ago, and the heat energy produced also re-melted the Earth's surface (Stevenson and Halliday 2014). Nevertheless, samples collected from the Moon by the lunar missions have been informative for dating the age of the Earth because the Moon's surface has undergone fewer changes over time. The 1971 Apollo 14 mission collected surface rocks and minerals which included samples representative of the early history of the Moon. Zircon mineral fragments have been dated to a minimum age of 4.5 billion years, which means that the Earth itself is older than this (Barbonil et al. 2017). On Earth, the oldest know minerals are tiny zircons isolated from sandstone rocks in Australia. One of these has been dated by highly accurate methods to 4.374 ± 0.006 billion years old, proving that the Earth had a solid crust by that time (Valley et al. 2014). Another striking example of the dating of ancient rocks is an analysis of the 4.02 billion-years-old Idiwhaa gneiss of Canada (Johnson et al. 2018), which was formed by partial melting of pre-existing rocks after a meteorite hit the Earth.

It can be assumed that the Earth contained no life at the time of its formation. It is thought that after cooling of the Earth's crust and the emergence of water, a "prebiotic" era began under the continuing influences of cosmic radiation and intense light from the Sun. This era, the time before biological life began on Earth, would have involved abiotic changes in surface rocks. These events may have led to the presence of chemical elements and more complex "building block" compounds that underwent chemical reactions to generate molecular precursors of life (Arndt and Nisbet 2012). There is also the possibility that organic (i.e., carbon-containing) material may have arrived in the atmosphere from extra-terrestrial sources, such as meteors or interstellar dust.

The first steps to life are thought to have occurred around 4 billion years ago and may have involved entities consisting of a few macromolecules or micrometre scale structures (see Sect. 3.2). This rough estimate is based on the assumption that the harsh conditions during the first 0.5 billion years after the formation of Earth would have been unsuitable for the establishment of life. Further time would then have been needed for the pre-biotic precursors undergo reactions and interactions that led to the origin of living cells. A separate school of thought, begun by Fred Hoyle and N Chandra Wickramasinghe (1978), is that living cells were delivered to the early Earth through impacts of comets or asteroids, a concept termed panspermia. This concept is not favoured by most biologists.

Fossils of precursors smaller than cells are extremely unlikely to be identifiable. However, geochemical methods have identified traces of carbon-based chemicals in rocks. These provide a signature that living organisms were present before 3.8 billion years ago (Mojzsis et al. 1996). The earliest known indications of living entities come from fossils discovered in Australian and Scottish rocks, which were dated to be between 3.8 and 3.4 billion years old by geological methods. The fossils themselves, formed from layers of ancient cyanobacteria cells, are known as stromolites (Lowe 1980). Possible stromolites older than 3.4 billion years have been reported from Greenland, but it has been disputed whether these are indeed fossils (Nutman et al. 2019).

The presence of deoxyribonucleic acid (DNA) also provides a molecular signature of life (we discuss DNA in Chap. 4). Although DNA has been recovered from samples thousands of years old, such as mammoths frozen in the permafrost (the oldest recovered DNA to date is from mammoths around 1 million years old (van der Valk et al. 2021), DNA cannot be recovered from fossil material of vast age. As a consequence, hypotheses about the very early forms of life are based on geochemical studies of rocks, models of how life processes may have developed and laboratory experiments that seek to replicate the likely conditions on the early Earth. Study of the geochemistry of the planet Mars may also yield hints on the prebiotic chemical environment of the early Earth, because the surface of Mars has retained very old rocks due to its less dynamic geological history (Farley et al. 2013). The models contain many interesting ideas about possible mechanisms, but since the experimental evidence is missing and the models are based on indirect evidence, they may diverge from the historical truth.

3.2 Prebiotic Chemical Constituents on the Earth Around 4.5 Billion Years Ago

Characteristics of the early atmosphere probably included heavy thunderstorms with large electric discharges. Carbon was present throughout the early Solar system and on the early Earth. The high energy provided by the discharges and/or meteor hits would have had a catalytic action that is thought to have enabled carbon (C), hydrogen (H), nitrogen (N), phosphorous (P) and other elements to combine into prebiotic compounds such as cyanide (CN). The atmosphere of the early Earth contained hydrogen (H_2), nitrogen (N_2), carbon dioxide (CO_2) and several other gases including possibly methane (CH_4) and ammonia (NH_3), but is thought to have been oxygen-free. With the cooling of the Earth's atmosphere and the development of a solid outer crust, liquid water would have begun to accumulate in depressions in the Earth's surface. These pools, in contact with the underlying rock and continuing to receive external energy from UV radiation or electrical storms, are thought to have provided environments suitable for additional chemical reactions; for example, the reaction of iron salts with hydrogen cyanide (HCN) to form ferrocyanide (Fig. 3.1) (Patel et al. 2015). A related proposal is that interactions occurring in water between organic, carbon-based molecules and the surfaces of rocks, minerals or clays would lead to clustering of molecules and would provide opportunities for additional chemical reactions.

Fig. 3.1 Model for chemical reactions following a meteor impact. It is proposed that the energy of the impact (impact shock) allowed production of atmospheric HCN. Chemical reactions of HCN with water and compounds from the rocks on Earth could lead to formation of ferrocyanide and mixed phosphate and chloride salts in water on the early Earth (From Patel et al. [2015], with permission from Springer-Nature)

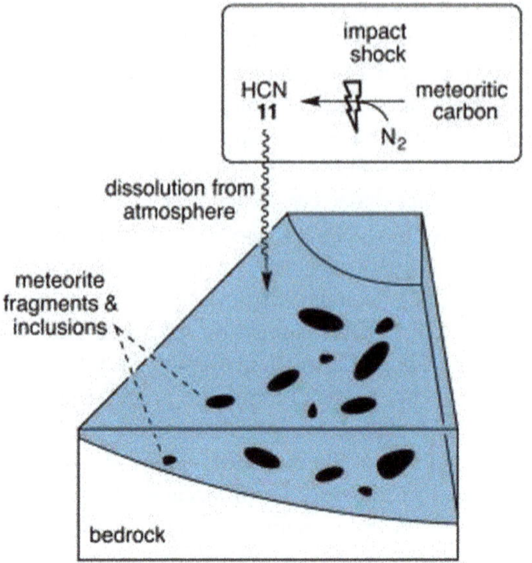

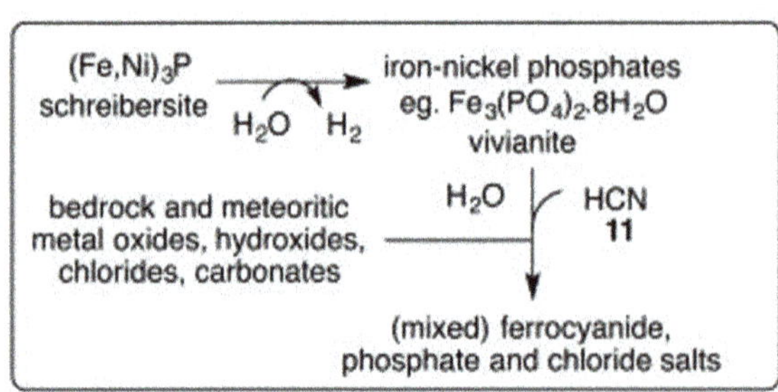

All living organisms are made up of molecules based predominantly on the element carbon. Why did life evolve as a carbon-based system? Carbon is one of the most abundant elements on Earth and its atoms are the smallest that have a valency of four: the importance of this valency is that both linear and branching chemical compounds can be formed. Unlike many other elements, carbon atoms will bond to each other, to form long, polymeric chains or ring structures, both of which permit the assembly of very complex molecules, such as DNA or polysaccharides. Carbon also readily forms chemical bonds with other abundant elements: hydrogen, nitrogen, oxygen, phosphorous and sulphur, and metals such as zinc or iron. Together with carbon, these elements form most of the mass of living cells.

Life on Earth takes part in, and depends on, a process called the carbon cycle, in which carbon atoms on Earth are in flux between rocks, oceans, the atmosphere, plants, soil, and fossil fuels. Living organisms take part in the so-called "fast carbon cycle" in which plants, phytoplankton and photosynthetic bacteria fix atmospheric CO_2: i.e., they drive the reaction of carbon dioxide and water to form sugars and oxygen. This enables them to synthesize a large number of organic compounds and nourish themselves through photosynthesis. The evolution of these organisms that add oxygen to the atmosphere allowed for the evolution of other organisms (termed heterotrophs; such as animals), that could utilise the atmospheric oxygen as an energetic component in metabolic pathways. Heterotrophs depend on photosynthetic organisms or other heterotrophs for their nutrition. By breaking down the sugars of plants, heterotrophs ultimately release CO_2 back to the atmosphere. Because of the high chemical reactivity of carbon-containing compounds, a very large number of synthetic and degradation pathways have evolved in living organisms (Rauchfuss and Mitchell 2008).

How did life begin from carbon-based compounds? In 1871 Charles Darwin speculated in a letter to Joseph Hooker (https://www.darwinproject.ac.uk/letter/DCP-LETT-7471.xml) that life might have begun in warm ponds. Models of reactions of carbon-based chemicals in water (the "primeval soup") were proposed by Alexander Oparin (1924) and John Haldane (1929), and were later termed the Oparin-Haldene hypothesis. Following a suggestion from Harold Urey that the lack of oxygen on the early Earth would have helped to promote the stability of carbon compounds, Stanley Miller devised an experimental test in which he boiled 100 ml of water in a flask (flask 1 in Fig. 3.2) and added 10 ml of H_2, 20 ml of CH_4 and 20 ml of NH_3 to a sealed glass apparatus. An electric discharge device (2) was operated to represent the effects of lightening strikes and produce free radicals. These triggered reactions in the gas phase in vessel 3 and any non-volatile products then accumulated in vessel 1. Importantly, 1 ml of a saturated mercury chloride solution was added to prevent

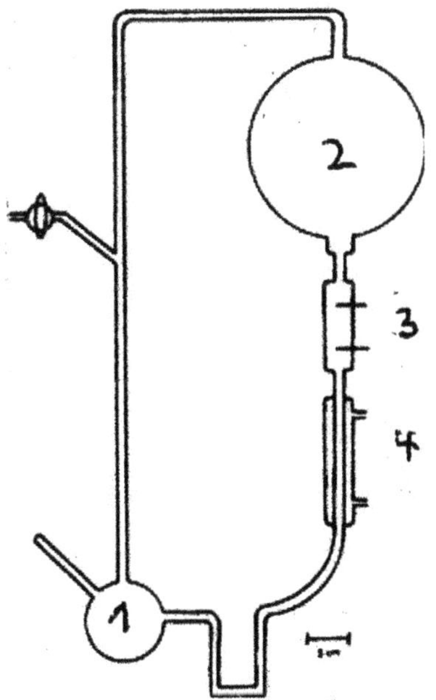

Fig. 3.2 Apparatus to investigate the products formed at 100 °C from water, hydrogen, methane and ammonia. The water is boiled in vessel 1. Water vapour and the gases react in vessel 2. Gases are exposed to electric discharge in device 3 and are refluxed by the cooler 4 (Copied with permission from S.L. Miller [1953])

the growth of living microorganisms. The experiment was run for a week and then the chemical products in vessel 1 were analysed.

Analysis by chromatography revealed that two different amino acids, alanine and glycine, had been formed and were mixed with various other organic compounds. These amino acids matched those found in the peptides and proteins of living organisms, although both L- and D-isomers were detected in the Miller experiments, whereas biological organisms specifically contain L-amino acids (Miller 1953).

The experiment of Miller gained a lot of attention. It was compared to purely statistical models for the creation of complex organisms by the association of atoms. For example Jacques Monod (1972) argued that the probability of formation of a living cell from atoms is comparable to the probability that a locomotive could form by itself in an iron mine. The chance of such an event is so low that it would not happen over billions of years. However, the situation is dramatically altered if the assembly can start from preformed building blocks such as amino acids. In Monod's scenario, the assembly of a "locomotive" (a living organism) from the parts of such a machine (the amino acids) would be much easier, even if many trials would still be needed to find the right assembly. The Miller experiment suggested how amino acids could have appeared as precursors to life through abiotic chemical reactions on the early Earth.

The Miller experiments have been repeated by other researchers, and also starting from different mixtures of prebiotic gases and reaction conditions. The simple chromatographic detection method was later replaced by mass spectroscopy. The results were confirmed and about 28 amino acids have been assayed (Ruiz-Mirazo et al. 2014). Micrometer scale soft, organic structures including a wide variety of proteinaceous and non-proteinaceous amino acids have been reported (Bassez et al. 2012). The results stimulated an extensive search for other prebiotic compounds of possible biological importance. Under varying conditions, synthesis of many relevant compounds including nucleotides was observed (for example, Ritson et al. 2018). The polymerization of nucleotides to form RNA molecules was reported (reviewed by Saladino et al. 2016; Šponer et al. 2016), and even the formation of lipids and membrane-like structures (De Souza et al. 2017). More recently, it has become more fully appreciated that tons of organic material are delivered to the Earth every day from space (Bernstein 2006). All these data indicate that a surprisingly high variety of organic compounds might have existed in prebiotic times, including compounds similar to components of known biological molecules. The information content contained in the chemistry and structure of these organic compounds is thought to have enabled the formation and function of precursors to living organisms, and has been named Chemical Evolution (Rauchfuss 2005; Rauchfuss and Mitchell 2008). The books by Rauchfuss review a much larger selection of hypothetical visions on how prebiotic chemistry could connect with the origin of life than are presented here.

3.3 Models for Molecular Precursors of Life

Cells are the fundamental unit of life and yet are already complex entities. Cells have many constituents including an outer lipid-based membrane which partitions the living cell from the surrounding environment, protein molecules, and multi-molecular complexes. Cells require energy sources to drive metabolic processes, and have mechanisms for accurate replication of a genome composed of DNA and its transcription into RNA (discussed in Chap. 4). Evolution has resulted in two fundamentally different categories of cells: prokaryotic cells, which have a relatively simple internal organisation with all major processes taking place within the cytoplasm, and eukaryotic cells, which contain membrane-bound internal organelles (discussed further in Sect. 3.5). In eukaryotic cells, most DNA is compartmentalised within a nucleus. Other important organelles are mitochondria that produce the bioenergetic molecules adenosine triphosphate (ATP) and guanosine triphosphate (GTP). Mitochondria also contain their own DNA. In humans, the mitochondrial genome makes up ~1% of the total genome.

Many different models for the molecular precursors of life have been proposed, which have focused on different components and fundamental properties of a cell. Different groups of researchers have focused either on information transfer (DNA or RNA mechanisms), on energy and metabolism, or on the roles of lipids in compartmentalisation from the environment. Which of these could have been the "first step" to life has been hotly debated. In the last decade, more integrative views have started to emerge, particularly that different types of molecules could have been present at the same time period and so different attributes of cells would not necessarily have started sequentially. An important example related to prebiotic chemistry is the experimental demonstration that RNA, protein and lipid precursor molecules can all be obtained by reductive reactions of hydrogen cyanide (HCN) and its derivatives under energy provided by UV radiation (Patel et al. 2015). As discussed in Sect. 3.2, HCN, phosphate and cyanide salts are very likely to have been present on the early Earth. The authors stress that, for chemical reasons, it is unlikely that all these reactions could take place simultaneously. They envision that different reactions might have taken place in different environments (pools or streams) that later mingled together. The low random chance of this happening would accord with the origin of life being a very rare event.

Another important concept for the transition from prebiotic conditions to biological life is the inception of molecules with "self-assembly" properties. This property would have enabled increased local concentrations of chemicals through formation of molecular aggregates. This could have increased the avidity for molecules to bind to inorganic surfaces. An increased local concentration of a particular molecule would have made interactions with other types of chemicals energetically more favourable, potentially leading to new reactions, new chemical products and new sources of chemical energy. Molecular self-assembly is also fundamental to the formation of structures with a distinct inside segregated from the general environment (compartmentalisation). Once compartmentalisation was achieved, different surface and interior chemical environments would have become possible and could lead to different types of chemical reactions (Monnard and Walde 2015). Figure 3.3 from Monnard and Walde's article summarises several models of how different chemical events in the prebiotic era led ultimately to the emergence of living cells.

3.4   More Detailed Models of the Origin of Life

Section 3.3 outlined concepts that guide us to hypothetical models of how the precursors of cells were formed. This section deals with more detailed models that are based on molecular mechanisms involving autocatalysis or compartmentalisation or on back extrapolation by use of phylogenetic trees. The models are all of the "if-then" nature and describe possible mechanisms that may be connected to the origin of life. According to the principles of natural sciences (Popper, 1963, 1994), proof for a model can be obtained only by experimental evidence. However, as outlined above, direct experimental evidence about the origin of life is missing because traces of the earliest precursors are missing. Evidence could in principle be obtained if the origin of life could be achieved experimentally. However, even if this could be done, we would not be sure if this would represent the historical past events (i.e., could life originate by multiple mechanisms, or is there one unique mechanism that could potentially be replicated?). At present, researchers have developed a number of plausible models but the historical process remains unknown.

3.4.1   The RNA Hypothesis

All living organisms on Earth require an information storage system to allow for the replication of cells during the lifespan and for the characteristics of a particular species to be passed to the next generation. For all organisms that we know of today, the inheritable information is encoded in DNA, but this may not have been the case for life's precursors or the earliest forms of life. Many researchers have considered that ribonucleic acid (RNA) is a better candidate molecule for the first form of information storage (Gilbert 1986). In contrast to DNA, RNA has a non-linear structure with many branches for interactions with other molecules. It not only stores and replicates genetic information but also mediates several other functions, the best known of which is the transfer of an amino acid to a nascent polypeptide by a transfer-RNA molecule. However, RNA is less chemically stable than DNA and the replication of RNA is less accurate than that of DNA. The latter is a disadvantage for accurate transfer of information, yet inaccurate replication might have been advantageous during the earliest stages of life, to enable the development of diverse systems. The present and possible past functions of RNA have been reviewed by Nelson

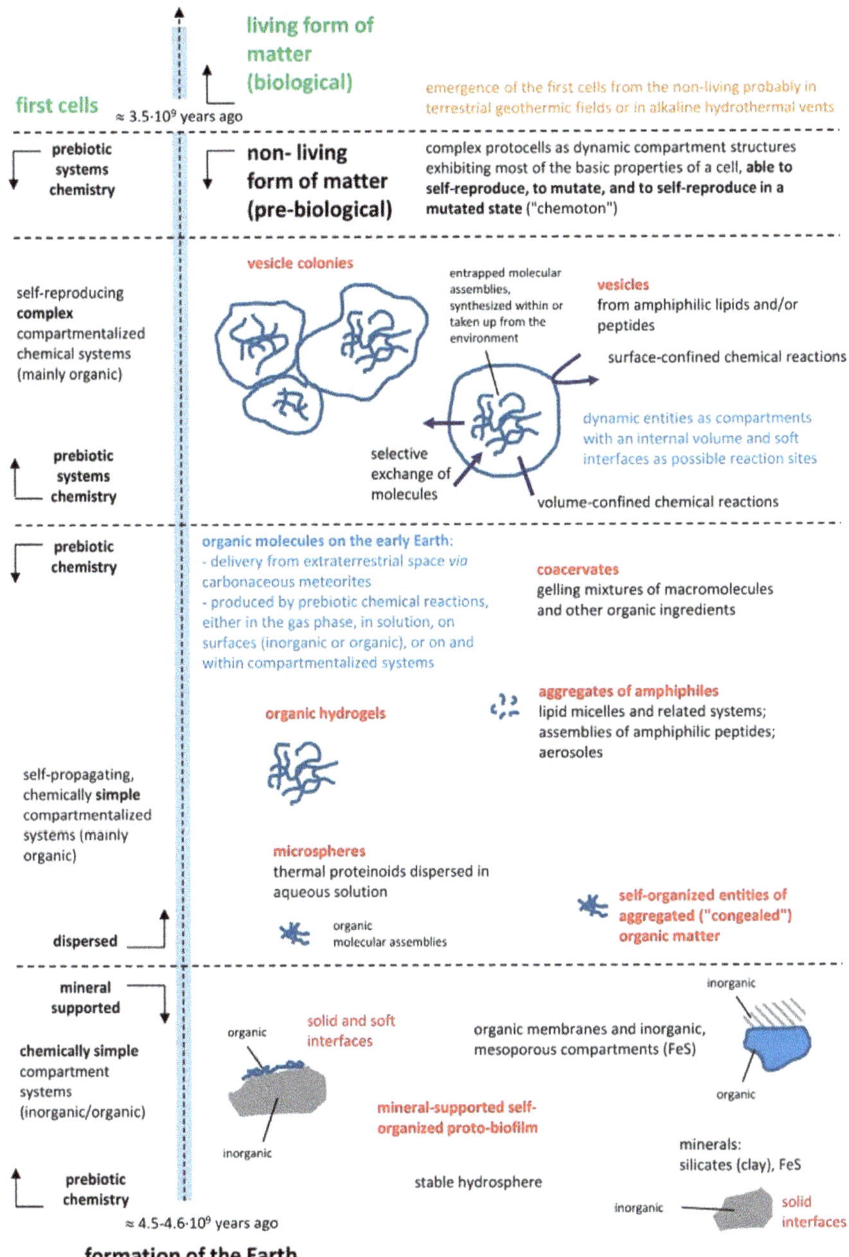

Fig. 3.3 Summary of the possible chemical steps that led to the beginning of life. The model assumes terrestrial sources of the initial organic compounds (From Monnard and Walde [2015], Life [Basel] [Open access, Attribution 4.0 International (CC BY 4.0)])

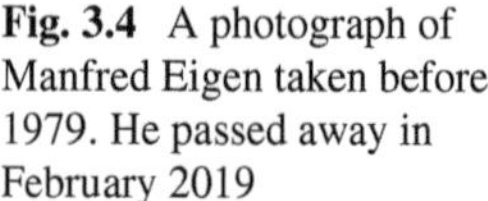

Fig. 3.4 A photograph of Manfred Eigen taken before 1979. He passed away in February 2019

and Breaker (2017). Although the model of an "RNA world" that preceded life has become dominant in recent years, it has remained unclear how RNA could have given rise to DNA prior to the existence of the type of enzymes that are used by cells to generate RNA transcripts from the DNA template. It has also been considered that RNA and DNA may not necessarily have had sequential origins. A recent study (Xu et al. 2019), showed that a nucleotide base-like compound, 2′-deoxy-2-thiouridine, could be synthesised under prebiotic-like conditions from a chemical intermediate of prebiotic RNA synthesis. Under the same conditions, 2′-deoxy-2-thiouridine could be converted to 2′-deoxyadenosine, one of the building block for DNA. These results have led to a new model, that RNA and DNA might both have been synthesised from the same precursors in the pre-biotic world. According to this model, only later did RNA and DNA adopt very different functions in living cells as a result of their different chemical properties and stabilities.

With regard to the "RNA world" model, there are several types of RNA in modern cells that serve different purposes. Manfred Eigen (Fig. 3.4) selected a transfer-RNA (t-RNA, specifically the t-RNA of phenylalanine) as the potential primordial "gene" (named Urgen by Eigen [1979]). This choice was made because he favoured its small size (76 nucleotides), its high conservation (there are only two sequence changes in this tRNA between mammals and *Drosophila* flies), its mirror image symmetry and high stability.

Later, Eigen and Winkler-Oswatitsch (1981) calculated a phylogenetic tree for the t-RNA of phenylalanine (Fig. 3.5). Phylogeny is a widely used computational method (Felsenstein 1996). Phylogenetic trees of biological relationships are constructed based on properties of organisms that change with evolution: for example, the

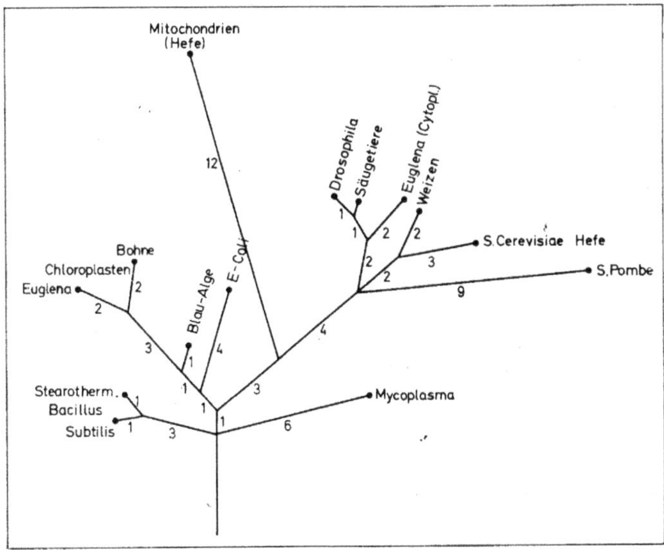

Fig. 3.5 Phylogenetic tree of t-RNA for phenylalanine, based on sequences from 14 organisms. (Reproduced with permission from Eigen, 1979)

sequence of nucleotides in DNA or RNA, or of amino acids in proteins. The trees make useful comparisons between modern organisms. With certain assumptions, the may also allow "back tracing" from existing modern sequences to hypothetical ancestral sequences. Other indirect assumptions are drawn from knowledge of biological mechanisms in modern organisms.

For many gene sequences, the information about the primordial sequence is almost impossible to infer, because of high sequence divergence between species. However, because of its high conservation, a phylogenetic analysis of the t-RNA could be carried out. The results suggested that the "Urgen" would have been closely related to the sequences of t-RNA for phenylalanine in modern *Bacillus subtilis* bacteria (Fig. 3.5) (Eigen and Winkler-Oswatitsch 1981). Between 1981 and the present day, many more t-RNAs of different species have been sequenced and it appears time to re-investigate the phylogeny of t-RNA. Problems in the interpretation of the t-RNA phylogeny (Gesell and Schuster 2014) should also be considered. In conclusion, the idea of t-RNA as a primordial gene as proposed by Eigen and other authors should be taken as a general concept and not restricted to a specific t-RNA. The concept of a primordial gene was later extended to a model of an autocatalytic hyper-cycle acting as a first self-replicating, self-organizing entity (Eigen and Schuster 1979; Eigen 1992).

3.4.2 *The Hyper-Cycle Model*

Based on the properties of all RNAs known today, the hypothetical Urgen of Eigen was hypothesised to replicate efficiently only in a process catalysed by enzymes; in living organisms these correspond to replicases. Eigen conceived a circular kinetic reaction (Fig. 3.6) named the hyper-cycle, in which both the RNA and the replicase become optimized in an autocatalytic reaction. As a result a system will develop which runs at higher rates than competing reactions and will finally stabilize in an optimal state. In Fig. 3.6 the blue circles labelled I1 to I5 in the centre indicate the replication cycles of the RNA. Each I with a different index number represents RNA containing small sequence variations caused by small replication errors. The green arrows indicate the formation of a replicase protein based on the sequence of I. Depending on the exact sequence of I, slightly different versions of the replicase are formed (E1 to En). In turn, the replicases act catalytically on the rate of replication of I (black arrows in Fig. 3.6).

The groups of Eigen and Schuster (1979) selected a cyclic reaction for their model because this type of reaction has a higher efficiency than a linear reaction. They were also stimulated by cyclic reactions of known biological importance, such as the Krebs cycle and citric acid cycle. As for all biological reactions, the hyper-cycle would run under irreversible conditions in which the reactants are supplied in excess.

Eigen and Schuster solved the rate equations for the hyper-cycle under a number of relevant conditions. Autocatalysis was found to be effective only under irreversible conditions. This catalysis may speed up RNA replication and the production of the enzyme E. Replication error mutations would alter the sequence of the RNA: the existence of RNA variants is essential for any selection mechanism. The model postulates that hyper-cycles of the different RNAs would run at different speeds and produce different proteins (E1-En). Under given environmental conditions, the hyper-cycles of the variant RNAs would compete with each other. The fastest or otherwise most advantageous system would come to dominate, providing a form of selection. According to discussions by Eigen and Schuster (1979) and Szostak (2017), other important properties of the hyper-cycle are a once-for-ever selection under given conditions, the utilization of small selective advantages and a selection against parasitic reactions. Eigen and Schuster have also discussed the favourable effect that compartmentalization would have, either in the form of a virus-like structure or a surrounding lipid membrane.

A limitation of this model is the requirement for a structured and functional protein, the enzyme. Although laboratory experiments indicate that amino acids can be formed under pre-biotic conditions (Sect. 3.2), it remains unclear how these would polymerise to form a correctly folded, functional enzyme. In living organisms, DNA encodes RNA, which encodes proteins, but this synthetic route would not have been available in the pre-biotic world. An alternative scenario is that a non-enzymatic mechanism for copying RNA from a template existed in the prebiotic world. Progress has been made in demonstrating mechanistic steps of RNA copying under laboratory conditions (Zhou et al. 2019).

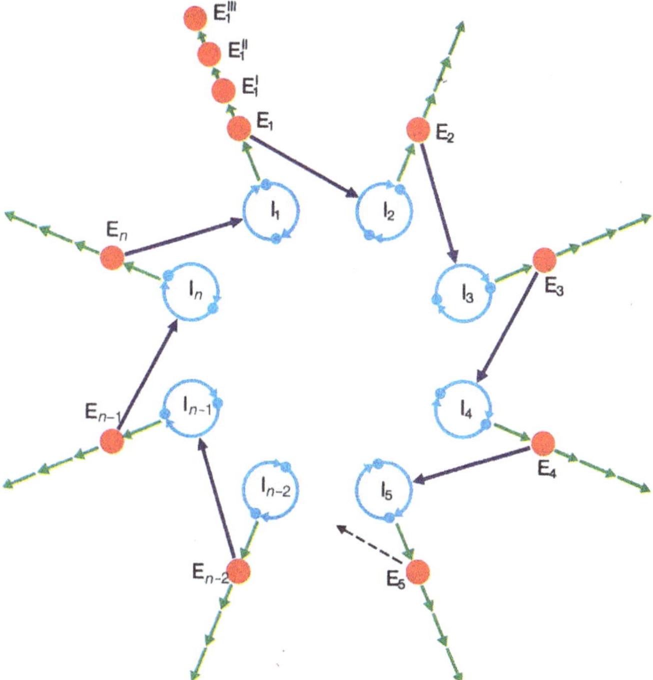

Fig. 3.6 Hypothetical autocatalytic hypercycle according to Eigen and Schuster (1979). Blue circles stand for RNA-replication and I for RNA. E stands for hypothetical replicases which are translated from the RNA and catalyse its replication. I and E may be altered by replication errors: different species are indicated by indices. Under irreversible conditions the circle will produce the protein E and RNA with a modified sequence, providing a mechanism for mutation and generation of variants to be acted on by selection according to environmental conditions. Copied with permission from Eigen (1992)

3.4.3   The "Metabolism First" Model

As we discussed in Sect. 3.2, chemical reactions on the early Earth depended on external sources of energy, such as UV radiation or electrical discharges from lightening. In contrast, living cells are self-sustaining systems that support themselves through the gathering of energy through cyclic chemical reactions that then maintain their organisation and functions. These processes are referred to as cell metabolism. Adenosine triphosphate (ATP) is a universal bioenergetic molecule and, in all forms of life, gradients of protons (H^+ ions) across membranes are used to generate energy to drive ATP synthesis (Lane et al. 2010). Given the central importance of metabolism to maintain life, some researchers consider that metabolic pathways must have originated before the property of information replication by DNA could be set up. The "metabolism first" model proposes that suitable conditions could have occurred in the chemical-rich and hot environments of hydrothermal vents. Such vents, located on the ocean floor, are equivalent to hot springs or geysers on land in that they expel geothermally-heated water (ranging from 50C to 400C). Hydrothermal vents were first discovered in 1977 (Corliss and Ballard 1977) as a new type of ecosystem with high temperature and high pressure, rich in minerals and with limited or no light. Despite these apparent challenges, many unique species of living organisms are found at vents. These include bacteria capable of surviving at high temperatures, which convert minerals emitted from the vents into energy-rich compounds that can then be used by other organisms. This conversion process is termed chemosynthesis. Hydrothermal vents are known to have been present on the early Earth and their unique physicochemical properties, in which hydrogen and carbon dioxide interact to generate carbon compounds such as methane by abiotic processes, have attracted scientific attention to the possibility that vents might have been the sites at which certain metabolic reactions originated.

The geochemical reaction pathways at vents are most similar to biochemical pathways of acetogenesis and methanogenesis, which are characteristic of certain modern microorganisms that live anaerobically (Martin et al. 2008). Alkaline (pH 9–11) hydrothermal vents, which are found on the modern mid-Atlantic ocean floor, are especially interesting because they emit hydrogen-rich material over geological timescales, a process that results in natural proton gradients. It has been suggested that these gradients within the physically confined vent could have provided the initial conditions under which proteins could harness the energy of a proton gradient (i.e., before transporter proteins had evolved to set up ion gradients across the lipid bilayer membrane of living cells). Access to the energy source of a proton gradient within the confines of a vent would have been advantageous to permit energetically unfavourable chemical reactions to proceed (Martin et al. 2008). However, other researchers have questioned the feasibility of submarine vents as a site for prebiotic origins, because laboratory experiments have shown that wet-dry cycles (rather than constant aqueous conditions) provide more favourable conditions for polymerisation of amino acids or ribonucleotides (Higgs 2016).

3.4.4 The Proto-cell Model

The RNA world models discussed in Sects. 3.4.1 and 3.4.2 focus on chemical reactions that can be carried out in test tubes with minimal components. However, from observations of biological specimens by microscopy it has been known from the seventeenth century onwards that all living organisms are composed of cells. The majority of organisms consist of a single cell, whereas larger organisms (principally animals, land plants and some fungi) are multi-cellular: they contain many cells that are stably organised into tissues. In all cases, every cell is segregated from the environment by a lipid-based cell membrane. This means that, whatever chemical reactions supported the appearance of prebiotic chemicals, compartmentalisation of these reactions away from the general environment must have been another essential step to cell-based life.

Considering the central role of compartmentalisation for the functioning of cells, Nobel prize winner Jack Szostak (2017) has given much thought to the ability of lipids and fatty acids to form cell-shaped vesicles. He and his colleagues have postulated that the self-assembly of these molecules could have been the primary starting point for the first living cells, allowing for an internal compartment shielded away from the external environment. Blain and Szostak (2014) have proposed that the first living cells emerged from a population of vesicles that contained oligonucleotides, which they term "proto-cells", i.e. simple, cell-like entities with a lipid outer membrane and an internal space filled with genetic material. They have also performed laboratory experiments in which vesicle assembly is promoted by exposure to clay particles, or in which RNA is polymerised from oligonucleotides and replicated inside phospholipid or fatty acid-based vesicles. They have identified conditions under which proto-cells will divide and replicate (Joyce and Szostak 2018). In this compartmentalisation model for the origin of life, the formation of proto-cells and uptake, assembly and replication of RNA are not separate events, but processes that could have occurred concurrently. The authors are extremely optimistic that future experimentation may yield new life forms distinct from existing biological life.

3.5 The Model for a Last Universal Common Ancestor of Cells (LUCA)

Although all living organisms are formed of one or more cells, not all cells are equivalent. Organisms are classified as prokaryotes or eukaryotes depending on whether their constituent cell(s) have a relatively simple internal organisation without a separate nucleus (prokaryotes), or a complex internal organisation including multiple types of membrane-bound organelles and nucleus (eukaryotes). The prokaryotic cell is characteristic of two of the three domains of life, the Eubacteria and the Archaea (Fig. 3.7). Phylogenetic trees and the fossil record show that Eubacteria and Archaea correspond to the earliest types of cells to emerge on Earth (see also Sect. 3.1). The third domain of life, the Eukarya, that comprises organisms with more complex, nucleated cells, evolved much later, perhaps around 1.8 million years ago. In 1967, Lynn Margolis was the first to put forward a coherent model that mitochondria, (organelles of eukaryotic cells with a central role in energy transfer), could have originated from endosymbiosis of an α-proterobacteria by an archaeal cell. She also proposed that the photosynthetic plastids of plant cells arose by endosymbiosis. These models are broadly accepted today (Gray 2017). Although phylogenetic trees place eukaryotes as more closely related to the Archaea, it remains uncertain if the primordial eukaryotic cell derived directly from an archaeal cell, or from more complex cell fusions or gene transfers between archaeal and bacterial cells (Koonin 2010). This is another aspect of the evolution of life for which there is no direct historical evidence.

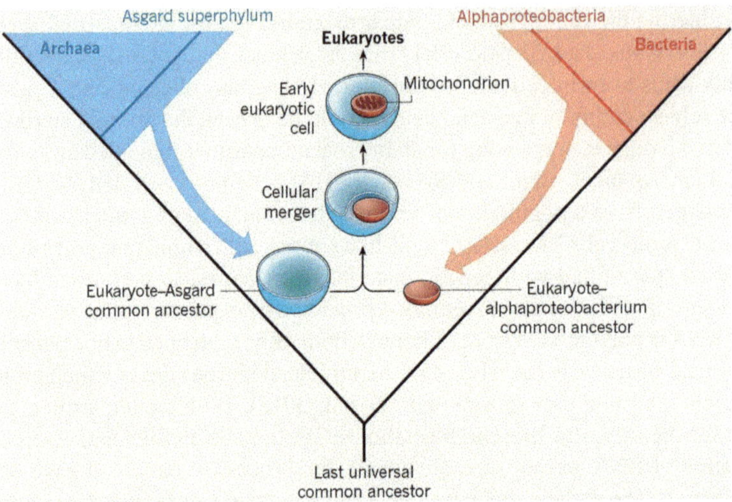

Fig. 3.7 Possible origins of the eukaryotic cell. The Asgard Archaea are from hydrothermal vents and include multiple genes conserved with eukaryotes (From McInerney and O'Connell 2017, with permission from SpringerNature)

Later, with the benefit of knowledge of the genome sequences of modern Archaea and bacteria, Weiss et al. (2016) took a computational approach to predict the properties of the first living cells. By applying phylogenetic methods, they could use the genomic sequence information of present-day species to identify the gene sequences that are most conserved between the modern species. By back extrapolation, these sequences become candidates for genes present in a primordial cell, termed the last universal common ancestor (LUCA) of bacteria and Archaea. The time at which this hypothetical cell lived remains undefined.

In this approach, 6.1 million protein-coding sequences from 1930 sequenced prokaryote genomes were sorted into 13 archaeal and 23 bacterial groups according to the taxonomic orders of the National Center for Biotechnology Information, USA. Protein families were sorted by Markov chain clustering (Enright et al. 2002) applying a program for correct sequence alignment (Landan and Grauer 2007). A gene family was considered to be present in LUCA if the given family was represented in at least two archaeal groups and two bacterial groups. Furthermore, archaea and bacteria were each required to be monophyletic (i.e., forming a single clade) in the corresponding phylogenetic tree. The phylogenetic trees were then calculated with the algorithm of Stamatakis (2014). By computational annotation of putative functions, 355 gene clusters that fulfilled the above criteria were identified and grouped into 21 functional categories (Fig. 3.8).

From these analyses the authors could develop a model for the LUCA cell and what its functional capacities could have been. The model proposes the LUCA cell to have been surrounded by a lipid membrane spanned by many proteins with channel or transporter functions, thus appearing like a simplified textbook diagram of a

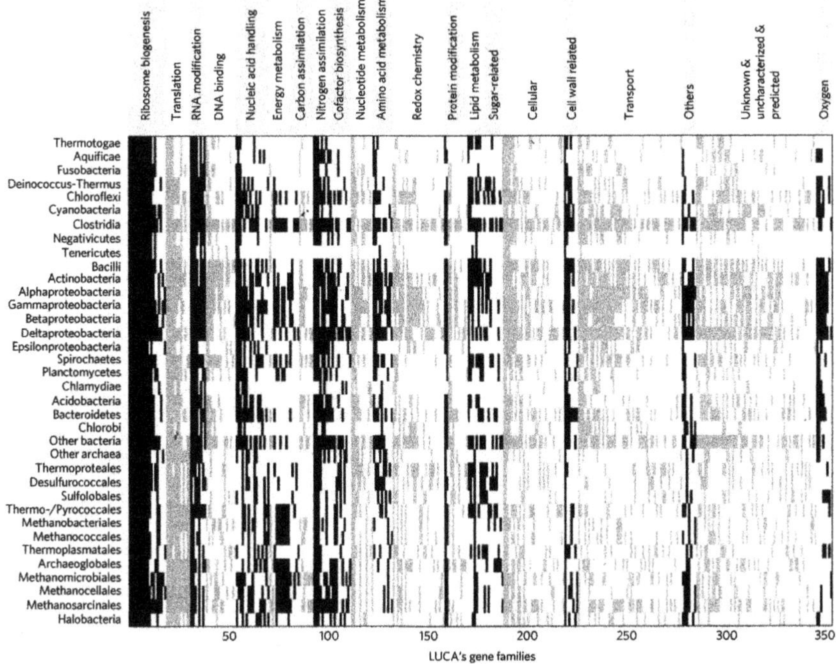

Fig. 3.8 The predicted genes of LUCA, grouped by functional categories obtained by computer annotation (top). Black or grey marks in the columns indicate the presence of a gene in the groups shown at the left. White marks indicate the absence of a gene (From Weiss et al. 2016)

cell without internal compartmentalization. The predicted gene products include molecules that capture the energy of a proton gradient, but not the molecules that would set up the gradient. For energy production, LUCA is predicted to have used the Wood-Ljungdahl pathway, in which hydrogen is used as an electron donor and carbon dioxide as the electron acceptor to synthesis acetyl-CoA and fix carbon under anaerobic conditions. This pathway is characteristic of some modern bacteria and Archaea called acetogens. The model depicts LUCA as a strictly anaerobic, H_2-dependent, thermophilic autotroph. This means that it was able to nourish itself (autotrophy) and would have been adapted to live at high temperatures: modern thermophilic bacteria live at temperatures between 41 °C and 122 °C. The authors speculated that this type of cell could have originated in hydrothermal vents. In view of the high complexity of LUCA it seems plausible that simpler precursor cells may have existed which may escape detection by the pathways of back extrapolation.

A perturbing result from this analysis was that the LUCA gene set included five oxygen-dependent enzymes, which may correspond to detoxifying enzymes (last column in Fig. 3.8). In view that the Earth's atmosphere would have been essentially oxygen-free at the likely time of LUCA, this finding may be erroneous. Weiss et al. (2016) discussed the inclusion of oxygen-related genes as possibly resulting from horizontal gene transfer from oxygen-dependent bacteria that would have existed much later than the time of LUCA. This finding might also be an artefact of the assumptions in the phylogenetic algorithms: these issues can have severe effects on the results obtained. Possible problems connected with the interpretation of phylogenetic plots computed from sequence data will be discussed further in Chap. 4.

More recently, these authors examined encoding of genes for metabolic networks in modern anaerobic bacteria to model properties of the last bacterial common ancestor. The results indicate an ancestor that could carry out gluconeogenesis and most similar in metabolism to modern thermophilic species of *Clostridia* (Xavier et al. 2021)

Another approach to the early evolution of cells based on phylogenetic methods has attempted to date LUCA and to gain insight into the emergence of the complex eukaryotic cell. In this study, phylogenetic trees were prepared based on 29 highly-conserved, universal proteins including proteins of the ribosome, and were calibrated to a timeline based on fossil, biomarker and geochemical data (Betts et al. 2018). The final tree was constrained by placing the maximum time for the root of the tree in the Hadean Eon (4.6–4 billion years ago). The reasoning for this was that the asteroid hit about 4.6 billion years ago that led to the formation of the Moon (Jacobson et al. 2014) would have wiped out any previous life forms. The authors tested the robustness of their outputs by applying different models of molecular evolution and testing the effects of different taxonomical sampling on tree topology. Overall, the emergence of LUCA was placed at >3.9 billion years ago, thus predating the end of the period of heavy bombardment of the Earth. Emergence of the primary kingdoms of life, the Eubacteria and the Archaea, was placed at <3.4 billion years ago and the emergence of the Eukaryota much later at <1.84 billion years ago. This model thus suggests a relatively long timeline from life's origin to the emergence of the three kingdoms of life that are present today; the Archaea, Eubacteria and Eukarya (Betts et al. 2018). Indeed, geological evidence indicates that a major rise in atmospheric oxygen, the "Great Oxidation Event", took place between 2.45 and 2.3 million years ago, there is also evidence that more local or transient oxygenation may have taken place from 3 billion years ago onwards. Thus, the first organisms capable of photosynthesis (see Sect. 3.2) must have evolved before this time period (Farquhar et al. 2011). One proposal is that communities of micro-organisms on land surfaces or in soil—such as mats of cyanobacteria—provided local oxygen from oxygenic photosynthesis long before the emergence of plants and the global rise in atmospheric oxygen (Lalonde and Konhauser 2015).

3.6 Common and Different Features of the Models

As discussed in Sect. 3.4, the "RNA world" model has gained support from laboratory experiments. However the metabolism-first model (Sect. 3.4.3) is not yet supported by laboratory experimental evidence. The hyper-cycle proposed by Eigen and his group described a rather simple reaction mechanism that is suggested to have been important at the very beginning of life. A very important feature of the hyper-cycle model is the autocatalysis by which the reaction may escape from many competing but slow mechanisms. The rate equations resemble those developed by Glansdorff and Prigogine (1971) for irreversible pathways and exhibit the formation of dissipative structures (complex structures generated by irreversible processes). As mentioned already, the Eigen group noted that compartmentalization would be favourable for the hyper-cycle, without being specific about how this feature would be set up. Over the last decades, considerable debate has continued on "RNA world" or "metabolism first" models for the origin of life, yet compartmentalisation would be required in either case.

In the work of Szostak and colleagues, compartmentalization in proto-cells is considered to be an essential circumstance that drove the origin of life. Their models are stimulated by the fascinating self-assembly properties of lipids and fatty acids. The pioneers of this model have performed experiments in which nucleotides were shown to polymerize to RNA (as the proposed information carrier) within vesicles.

The more recent predictions of the gene content of LUCA cells has involved a completely different approach, back extrapolation from modern bacterial and archaeal genomes to predict the putative gene content of a precursor cell. These studies depend on the widely-used approach of construction of phylogenetic trees. However, these methods include the assumption that mutation rates on the early Earth were equivalent to the rates observed today. This is unlikely in view of the much thinner atmosphere on the early Earth. The inclusion of predicted genes for oxygen-dependent enzymes for a LUCA that would have existed at a time when there was no or little oxygen in the atmosphere suggests that further refinement of this approach is needed.

All three models are based on the properties and reactions of organic chemistry. Based on laboratory experiments, it is considered feasible that molecules such as the nucleotides that build RNA or DNA; lipids; amino acids, or precursors of these, could have come to be present on Earth at the prebiotic time. RNA or DNA replication with reading errors, enzyme catalysis and rate equations all follow the laws of chemistry. Formation of cell walls from lipids and fatty acids are also chemical self-assembly processes. The common features of all models suggest some important conclusions that are listed below. More examples can be found in the books (Rauchfuss 2005; Rauchfuss and Mitchell 2008).

1. A large fraction of the information content needed for the mechanisms of life comes from the laws of organic chemistry. Karl Popper (1994) used the phrase: life is reduced to organic chemistry.
2. Although inorganic compounds such as silica, calcium phosphates, iron-based and strontium-based compounds can be incorporated into living organisms, bio-mineralization processes are always under the control of organic biomolecules (Engel 2017). No organism composed solely of inorganic compounds has appeared during the billions of years of biological evolution on Earth. We discussed in Sect. 3.2 some of the advantageous properties of carbon chemistry. Inorganic chemistry, which means carbon-free chemistry, does not offer

lipid-like molecules for the construction of cell walls and cell membranes. Inorganic molecules with an information storage capacity and ability to replicate comparable to that of DNA and RNA are unknown. It remains a debatable question whether life can originate only from carbon chemistry, or if life forms based on inorganic elements could be feasible on other planets. In later chapters, we will discuss ideas about inorganic or "hybrid" living species in relation to artificial intelligence and the future of life (Chap. 9). We will also discuss current ideas about the possibility of life on other planets and the search for life on Mars (Chap. 10).

3. A feature not explicitly mentioned in the models is the crowding of molecules that occurs in cells and vesicles (Zhou et al. 2008). In the cytosol, concentrations and free energies of macromolecules are extremely high. Thus, interactions between different partners, which would not be possible under dilute conditions, are possible in compartments. For this reason, a very large diversity of interactions has to be expected under crowded conditions. This property may have helped in the initial steps of evolution of living cells once compartmentalisation from the environment had been achieved.

From these diverse models for the origin of life we have outlined several concepts for how it may have happened. A key question that remains is: Was life initiated only once by a single mechanism? Processes that occur only once cannot be investigated (Monod 1972), and thus cannot lead to predictions. It is plausible that the steps leading to the origin of life had a low but finite probability. This would agree with the generally accepted idea of astronomers that life could also originate on other planets providing that the chemical conditions are met (see Chap. 10). If this is true, it should be possible to set up sterile conditions to perform experiments here on our Earth to initiate new life.

References

Arndt NT, Nisbet EG (2012) Processes on the young earth and the habitats of early life. Annu Rev Earth Planet Sci 40:521–549

Barbonil M, Boehnke P, Keller B, Kohl IE, Schoene B, Young ED, McKeegan KD (2017) Early formation of the Moon 4.51 billion years ago. Sci Adv 3(1):e1602365. https://doi.org/10.1126/sciadv.1602365

Bassez MP, Takano Y, Kobayashi K (2012) Prebionic organic microstructures. Org Life Evo Biosp 42:307–316

Bernstein M (2006) Prebiotic materials from on and off the early earth. Philos Trans R Soc Lond B Biol Sci 361:1689–1700

Betts HC, Puttick MN, Clark JW, Williams TA, Donoghue PCJ, Pisani D (2018) Integrated genomic and fossil evidence illuminates life's early evolution and eukaryote origin. Nat Ecol Evol 2:1556–1562

Blain CJ, Szostak JW (2014) Progress towards synthetic cells. Ann Rev Biochem 83:615–640

Corliss JB, Ballard RD (1977) Oases of life in the cold abyss. Nat Geogr 152:441–453

De Souza TP, Bossa GV, Stano P, Steiniger F, May S, Luisi PL, Fahr A (2017) Vesicle aggregates as a model for primitive cellular assemblies. Phys Chem Chem Phys 19:20082–20092. https://doi.org/10.1039/c7cp03751a

Eigen M (1979) Das Urgen. Nova Acta Leopoldina 243:5–40

Eigen M (1992) Steps towards life. Oxford University Press, pp 1–173

Eigen M, Schuster P (1979) The hypercycle: a principle of natural self-organization. Springer

Eigen M, Winkler-Oswatitsch R (1981) Naturwissenschaften 68:282–292

Engel J (2017) A critical survey of biomineralization. Springer Briefs in Applied Science

Enright AJ, Van Dongen S, Ouzounis CA (2002) An ancient algorithm for large scale detection of protein families. Nucleic Acid Res 30:1575–1584

Farley KA, Malespin C, Mahaffy P, Grotzinger JP, Vasconcelos PM, Milliken RE, Malin M, Edgett KS, Pavlov AA, Hurowitz JA, Grant JA, Miller HB, Arvidson R, Beegle L, Calef F, Conrad PG, Dietrich WE, Eigenbrode J, Gellert R, Gupta S, Hamilton V, Hassler DM, Lewis KW, McLennan SM, Ming D, Navarro-González R, Schwenzer SP, Steelel A, Stolper EM, Sumner DY, Vaniman D, Vasavada A, Williford K, Wimmer-Schweingruber RF, the MSL Science Team (2013) In situ radiometric and exposure age dating of the Martian surface. Science 343:1247166. https://doi.org/10.1126/science.1247166

Farquhar J, Zerkle AL, Bekker A (2011) Geological constraints on the origin of oxygenic photosynthesis. Photosynth Res 107:11–36. https://doi.org/10.1007/s11120-010-9594-0

Felsenstein J (1996) Inferring phylogenies from protein sequences by parsimony, distance, and likelihood methods. Methods Enzymol 266:418–427

Gesell T, Schuster P (2014) Phylogeny and evolution of RNA structure. In: Methods in molecular biology. Humana Press

Gilbert W (1986) Origin of life: The RNA world. Nature 319(6055):D618. Bibcode:1986Natur.319..618G. https://doi.org/10.1038/319618a0

Glansdorff P, Prigogine I (1971) Thermodynamics theory of structure, stability and fluctuations. Wiley-Interscience

Gray MW (2017) Lynn Margulis and the endosymbiont hypothesis: 50 years later. Mol Biol Cell 28:1285–1287. https://doi.org/10.1091/mbc.E16-07-0509

Haldane JBS 1929 Origin of life. Ration Annu 148:3–10

Higgs PG (2016) The effect of limited diffusion and wet-dry cycling on reversible polymerization reactions: implications for prebiotic synthesis of nucleic acids. Life (Basel) 6:24

Hoyle F, Wickramasinghe NC (1978) Life cloud. J.M. Dent & Sons, London

Jacobson SA, Morbidelli A, Raymond SN, O'Brien DP, Walsh KJ, Rubie DC (2014) Highly siderophile elements in Earth's mantle as a clock for the Moon-forming impact. Nature 508:84–87. https://doi.org/10.1038/nature13172pmid:24695310

Johnson TE, Gardiner NJ, Miljković K, Spencer CJ, Kirkland CL, Bland PA, Smithies H (2018) An impact melt origin for Earth's oldest known evolved rocks. Nat Geosci. https://doi.org/10.1038/s41561-018-0206-5

Joyce GF, Szostak JW (2018) Protocells and RNA self-replication. Cold Spring Harb Perspect Biol 10(9):pii: a034801

Koonin EV 2010 The origin and early evolution of eukaryotes in the light of phylogenomics. Genome Biol 11(5):209. https://doi.org/10.1186/gb-2010-11-5-209

Lalonde SV, Konhauser KO (2015) Benthic perspective on Earth's oldest evidence for oxygenic photosynthesis. Proc Natl Acad Sci USA 112(4):995–1000. https://doi.org/10.1073/pnas.1415718112

Landan G, Grauer D (2007) Heads or tails: a simple reliability check for multiple sequence alignment. Mol Biol Evol 24:1380–1383

Lane N, Allen JF Martin W 2010 How did LUCA make a living? Chemiosmosis in the origin of life. Bioessays 32:271–280. https://doi.org/10.1002/bies.200900131

Lowe DR (1980) Stromatolites 3,400 Myr old from the Archean of Western Australia. Nature 284:441–443

McInerney JO, O'Connell MJ (2017) Mind the gaps in cellular evolution. Nature 541:297–299. https://doi.org/10.1038/nature21113

Martin W, Baross J, Kelley D, Russell MJ (2008) Hydrothermal vents and the origin of life. Nat Rev Microbiol 6:805–814

Miller SL (1953) A production of amino acids under possible primitive earth conditions. Science 117:523–529

Mojzsis SJ, Arrhenius G, McKeegan KD, Harrison TM, Nutma AP, Friend CRL (1996) Evidence for life on Earth before 3,800 million years ago. Nature 384:55–59

Monod J (1972) Chance and necessity: an essay on the natural philosophy of modern biology. Vintage. ISBN-10 0394718259

Monnard PA, Walde P (2015) Current ideas about prebiological compartmentalization. Life (Basel) 5(2):1239–1263. https://doi.org/10.3390/life5021239

Nelson JW, Breaker RR (2017) The lost language of the RNA World. Sci Signal 10(483):pii: eaam8812. https://doi.org/10.1126/scisignal.aam8812

Nutman AP, Bennett VC, Friend CRL, Van Kranendonk MJ, Rothacker L, Chivas AR (2019) Cross-examining Earth's oldest stromatolites: seeing through the effects of heterogeneous deformation, metamorphism and metasomatism affecting Isua (Greenland) ~3700 Ma sedimentary rocks. Precambr Res 331:105347

Oparin AI 1924 Proiskhozhdenie zhizny (The origin of life, Ann. Synge. Trans.) In: JD Bernal (ed) The origin of life. Weidenfeld and Nicholson, London

Patel BH, Percivalle C, Ritson DJ, Duffy CD, Sutherland JD (2015) Common origins of RNA, protein and lipid precursors in a cyanosulfidic protometabolism. Nat Chem 7(4):301–307. https://doi.org/10.1038/nchem.2202

Peslier AH (2020) The origins of water. Science 369:1058. https://doi.org/10.1126/science.abc1338

Popper KR (1994) Das Leben ist Probleme lösen. Chapers 1 (Wissenschaftslehre in entwicklungstheoretischer und in logischer Sicht), Chapter 2 (Wissenschaftliche Reduktion und die essentielle Unvollständigkeit der Wissenschaft), Piper, München, Zürich

Popper KR (1963) Science: problems, aims, responsibilities. Fed Proc 22(961):1–27

Rauchfuss H (2005) Chemische Evolution und der Ursprung des Lebens. Springer

Rauchfuss H, Mitchell TN (2008) Chemical evolution and the origin of life. Springer

Ritson DJ, Battilocchio C, Ley SV, Sutherland JD (2018) Mimicking the surface and prebiotic chemistry of early Earth using flow chemistry. Nat Commun 9:1821. https://doi.org/10.1038/s41467-018-04147-2

Ruiz-Mirazo K, Briones C, de la Escosura A (2014) Prebiotic system chemistry: new perspectives for the origin of life. Chem Rev 114:285–366

Saladino R et al (2016) A global scale scenario for prebiotic chemistry: silica-based self-assembled mineral structures and formamids. Biochemistry 55:2806–2811

Šponer JE, Šponer J, Nováková O, Brabec V, Šedo O, Zdráhal Z, Costanzo G, Pino S, Saladino R, Di Mauro E (2016) Emergence of the first catalytic oligonucleotides in a formamide-based origin scenario. Chemistry 22:3572–3586

Stamatakis A (2014) RAxML version 8, a tool for phylogenetic analysis and post-analysis of large phylogenies. Bioinformatics 30:1312–1313

Stevenson DJ, Halliday AN (2014) The origin of the Moon. Philos Trans A Math Phys Eng Sci 372(2024):20140289. https://doi.org/10.1098/rsta.2014.0289

Szostak JW (2017) The narrow road from the deep past: in search of the chemistry of the origin of life. Angew Chem Int Ed 56:11037–11043

Valley JW, Cavosie AJ, Ushikubo T, Reinhard DA, Lawrence DF, Larson DJ, Clifton PH, Kelly TF, Wilde SA, Moser DE, Spicuzza MJ (2014) Hadean age for a post-magma-ocean zircon confirmed by atom-probe tomography. Nat Geosci 7:219–223

van der Valk T, Pečnerová P, Díez-del-Molino D et al (2021) Million-year-old DNA sheds light on the genomic history of mammoths. Nature. https://doi.org/10.1038/s41586-021-03224-9

Weiss MC, Sousa FL, Mrnjavac N, Neukirchen S, Roettger M, Nelson-Sathi S, Martin WF (2016) The physiology and habitat of the last universal common ancestor. Nature Microbiology 1:16116

Xavier JC, Gerhards RE, Wimmer JLE, Brueckner J, Tria FDK, Martin WF (2021) The metabolic network of the last bacterial common ancestor. Commun Biol 4(1):413. https://doi.org/10.1038/s42003-021-01918-4.

Xu J, Green NJ, Gibard C, Krishnamurthy R, Sutherland JD (2019) Prebiotic phosphorylation of 2-thiouridine provides either nucleotides or DNA building blocks via photoreduction. Nature Chemistry 11:457–462

Zhou HX, Rivas G, Minton AP (2008) Macromolecular crowding and confinement: biochemical, biophysical, and potential physiological consequences. Annu Rev Biophys 37:375–397

Zhou L, Kim SC, Ho KH, O'Flaherty DK, Giurgiu C, Wright TH, Szostak JW (2019) Non-enzymatic primer extension with strand displacement. Elife 8:pii: e51888. https://doi.org/10.7554/elife.51888

Chapter 4
Biological Evolution

We live in a world of tremendous biological diversity with millions of species in the prokaryotic and eukaryotic kingdoms of life. A recent estimate of the biomass of the different kingdoms of life puts plants as by far the most abundant category, comprising 450 Gt of carbon out of an estimated 550 Gt of carbon for all life on Earth. Bacteria and Archaea are estimated as the next most abundant (77 Gt carbon), followed by fungi (12 Gt), protists (4 Gt) and animals (2Gt) (Bar-On et al. 2018).

As discussed in Chap. 3, the process that led to the origin of life is considered to have begun more than four billion years ago. The focus of this Chapter is on the processes after that first step, by which the current diversity of species has come to exist. We now know that this process is evolution under natural selection, which takes place over time as an interaction between the inherited characteristics of a species and the properties of its environment. Throughout most of human cultural history it was generally believed that species were immutable (unchanging) and had been created separately when the world was first made either by gods or by mythic beings. There were also beliefs that the Earth was only a few thousand years old. In modern language, this form of belief is called creationism (see Chap. 2). These misconceptions prevented the development of ideas that life forms could change slowly over time and that new species could emerge. However, based on detailed observations of nature and geology during the eighteenth and nineteenth centuries, as increasing evidence of animal and plant fossils and better scientific knowledge of the Earth's geology were discovered, naturalists in western European countries began to have doubts about creationism.

4.1   Charles Darwin and Darwinism

The man who fundamentally changed our knowledge of living species was Charles Darwin (Fig. 4.1). Darwin was born in Shrewsbury, England on 12th February 1809. In his own words, he did not do well at school (Barlow 1958). In 1825, his father sent

J. C. Adams and J. Engel, *Life and Its Future*,
https://doi.org/10.1007/978-3-030-59075-8_4

Fig. 4.1 Charles Darwin, in a photograph taken in 1854 (From Wikipedia, English version 2020)

him to begin studies in medicine at Edinburgh University. After two years, Darwin decided against becoming a physician and his father suggested that he should enter the clergy. It was with this goal in mind that Darwin entered Cambridge University in 1828. During his time there, his greatest pleasure was in collecting beetles. He also met and formed a friendship with the Professor of Botany, John Henslow, and developed botanical interests and knowledge from their long walks (Barlow 1958). After graduation in 1831, Darwin returned home and began to study geology in Shropshire followed by a geological walking tour in Wales with Professor Sedgwick.

The course of Darwin's life was changed on receipt of a letter from Professor Henslow with the news that Captain Robert Fitzroy of the *HMS Beagle* was looking for an unpaid naturalist to take part in a voyage of surveying and scientific exploration. The main goal was to visit Terra del Fuego and "return home by the East Indies" over a period of about two years. Although Darwin's father was initially opposed to the idea, Darwin's uncle, Josiah Wedgewood, provided persuasive arguments and Darwin was able to take the post. The *HMS Beagle* departed from Plymouth on 27th December 1831 when Darwin was twenty-two and Fitzroy was twenty-six. In the event, the voyage took five years. *HMS Beagle* toured the east and west coasts of South America, the Galapagos Islands, Tahiti, the coasts of Australia and New Zealand and the Cape of Good Hope of Africa. This great opportunity to observe, collect, and study specimens of many different species from within their natural habitats and to gain first-hand knowledge of the geological contexts, transformed Darwin as a scientist. His journal about the voyage and the treks on land, published as "Voyage of the *Beagle*" (1839) demonstrates how carefully he collected and documented materials, his eye for detail and his wide-ranging curiosity about the places and people visited during the voyage. His geological studies benefited greatly from reading

Charles Lyell's "Principles of Geology" (published 1830–1833) and from first-hand experiences in Chile of a volcanic eruption and earthquake.

The visit on the Galapagos Islands also proved very significant for Darwin's later development of the theory of evolution. This chain of islands in the Pacific Ocean emerged through geological activity (which is on-going to this day). On some of the islands, Darwin viewed barren land covered by solidified lava; on others, there were signs of new colonisation by life and a striking diversity of species. These observations helped Darwin to consolidate his observations that the land itself is continually changing, which stimulated his thinking about species and their relationship to their natural habitat and how new land might be colonised. However, it was only after his return to England in 1836 and the more detailed examination of bird specimens collected separately on the islands by himself and Captain Fitzroy, that Darwin realised that many represented different species, distinct from those on the mainland of Ecuador. The identification of species specific to particular Islands provided more evidence for the idea that species become modified and optimally adapted to their environment over time.

In Darwin's own words "After my return to England, it appeared to me… by collecting all the facts which bore in any way on the variation of animals and plants under domestication and nature, some light might perhaps be thrown on the whole subject" (Barlow 1958). Darwin's massive book, published 1859, "On the Origin of Species by Means of Natural Selection, or the Preservation of Favoured Races in the Struggle for Life", in which he interpreted his observations of the natural world and formulated theories, proposed a new way of thinking about species. It is considered by most biologists to be the most important book in biological science.

Darwin's extensive studies put together evidence that the nature of organisms could change over time, drawing on the Beagle voyage as well as topics that he had researched during the writing of the book. He noted how well adapted each species is to survive in its particular habitat, and also that individuals within a species are not completely identical but show a spectrum of minute variations. He also investigated the selective breeding of farm animals to gain evidence for adaptations and their inheritance. To explain how organisms could have become so well-adapted to their environments, he developed a theory of natural selection. Because natural resources are finite and yet successful breeding of living organisms tends to expand their populations, there is a constant struggle for natural resources: food, water, and other features of a suitable environment. According to their fitness for survival under the environmental conditions (which stems from the variation between individuals), some organisms within a species will be more successful than others in breeding and thereby passing their specific traits onto the next generation. This constitutes the natural selection of successful variants: by the combination of statistical probability and selection determined by the local conditions such as food and other prerequisites of life, organisms evolve in the direction of higher "fitness" within their environment.

Darwin based his theories on careful and extensive observations of the natural world and its living organisms. The voyage of *HMS Beagle* gave him a unique opportunity to appreciate the diversity of species in many lands and their geographical ranges. In principle, the facts which he documented were in most cases accessible to anybody. Apparently, many others could not see these facts because of fixed dogmatic views. The anatomist and pioneer in modern education, Thomas Huxley, made an exclamation upon reading *The Origin of Species*: "How extremely stupid not to have thought of that!".

In later editions of *The Origin of Species*, Darwin lists about thirty predecessors of his theory. Among them are Lamarck, Geoffroy Saint-Hilaire, W.C. Wells and his own grandfather, Erasmus Darwin. Remarkably, the British naturalist Alfred Russel Wallace, (1823–1913), conceived the idea of natural selection independently of Darwin (https://academic.oup.com/biolinnean/pages/alfred_russel_wallace). Wallace had travelled extensively for eight years (1854–1862) in the Malay Archipelago and collected many specimens of birds, animals and plants. He had documented that the native species were striking different in the region to the East of Borneo. This "boundary" has become known as the Wallace Line and is now known to demarcate the boundary between species from two distinct geographical regions, the Asian region and the Australian region. Wallace had also started to consider how species arose. As presented in his paper of 1855, he argued "Every species has come into existence coincident both in time and space with a pre-existing closely allied species" (Wallace 1855). During his travels in 1858, whilst suffering from a high fever, Wallace had the inspiration that species changed over time because only the fittest individuals passed on their traits to the next generation: the same idea that Darwin had termed natural selection. Wallace wrote to Darwin, as a fellow naturalist, of his idea. At this time, Darwin was still completing his book. It was arranged by colleagues that both would put their ideas into a joint paper, which was then read by Charles Lyell and Joseph Hooker to the Linnean Society in London (Darwin and Wallace 1858). Darwin's book published in 1859 then made a detailed case for evolution under natural selection in a broad and convincing way.

The ideas of Darwin became adopted in socialism and politics and had widespread appraisal. Darwin remains the *bête noir* of religious fundamentalists. In some states of the USA, fundamentalists have demanded, with partial success, that the theory of biological evolution and religious belief in creationism should be covered with equal weight in school textbooks. As a result, a typical bookstore may have neighbouring shelves with books on the science of biological evolution or religious creationism. Darwin discussed the controversy between evolution and religious belief in several of his books. He himself did not break with religion, remained Christian, and was buried in Westminster Abbey on April 19, 1882.

Later, in the 1940s, with more knowledge of mechanisms of how traits are passed from generation to generation, evolutionary biologists including the great Ernst Mayr made a "Modern Synthesis" of Darwinism and genetics to provide a unified evolutionary theory that incorporated Mendelian mechanisms of inheritance (discussed below) with natural selection, to address the issue of how species originate. Mayr (1997) summarised the five major proposals that Darwin advanced in his book:

1. Organisms evolve steadily over time to adapt to their environment
2. Different kinds of organism all descended from a common ancestor
3. Species multiply over time
4. Biological evolution takes place through gradual change

5.   There are inheritable variations between the individuals of a species and the mechanism that drives evolution is the competition among vast numbers of individuals, within and between species, for limited natural resources (Darwin called this natural selection).

4.2   Darwinism in the Light of Molecular Biology

It follows from Darwin's books that he knew that a carrier of inheritance must exist, but he did not know the molecules involved or the mechanism. He also did not know what a protein is or the way in which its amino acid sequence is determined. It is very admirable that Darwin derived his detailed theory and perspective from global studies of living and fossil organisms without knowing the molecular basis.

At the time of Darwin's birth, none of the major biological macromolecules deoxyribonucleic acid (DNA), ribonucleic acid (RNA), proteins, or polysaccharides were known and so there were no concepts of the functions of these essential molecules of biology. Pioneers in this field were Gregor Mendel (1822–1884) and Friedrich Miescher (1844–1895).Mendel (Fig. 4.2) founded the field of genetics through his studies on breeding garden pea plants in St. Thomas' Abbey at Brunn (now Brno). By hybridising different strains through generations and keeping quantitative records on the appearance of the peas, Mendel developed three principles to explain the inheritance of distinct traits such as pod shape, or pea colour or smoothness. These experiments continue over eight years and Mendel published his results

Fig. 4.2 Photograph of Gregor Mendel about 1866 (From Wikipedia, English version 2020)

Fig. 4.3 Friedrich Miescher and the purification of nuclein.
Lefthand, Photograph of Miescher from about 1871 (from Wikipedia, English version 2020). Righthand, Glass tube contains salmon sperm DNA isolated by Miescher at the Univerity of Basel. Reproduced from Dahm 2005 (open access). The tube is the possession of the Interfacultary Institute for Biochemistry, University of Tübingen

in 1866 (https://www.nature.com/scitable/topicpage/gregor-mendel-a-private-scient ist-6618227/). His breakthrough publication was published in a relatively obscure journal and was ignored amongst the wider scientific community for about 30 years, until the paper came to the attention of the researchers Carl Correns, Hugo de Vries and Erich von Tschermak in 1900, and William Bateson in Cambridge in 1901 (Bateson 1901). Bateson translated Mendel's paper into English and is credited with inventing the term genetics. It is not known whether Darwin was aware of Mendel´s results because Mendel was working in isolation and "On the Origin of Species.." was published before Mendel's paper.

Freidrich Miescher (Fig. 4.3) was a Swiss physician and biologist. During the middle of the nineteenth century, the cell theory of life (that all forms of life consist of cells and that all cells arise from pre-existing cells) took shape through the research of Schwann, Virchow and others. In 1866, Ernst Haeckel proposed that the nucleus of cells contained the factors that transmit inherited characteristics. Miescher developed a strong interest in what was then called 'the foundations of life" (Dahm 2005). After qualifying as a physician in 1868, he turned to research, becoming the student of the biochemist Felix Hoppe-Seyler at the University of Tübingen. His goal was to identify the chemical nature of cells. To obtain cells in sufficient quantities for the analysis, he worked with leukocytes collected from pus on bandages from the nearby hospital. After initial extraction and identification of proteins and lipids, Miescher developed a procedure to isolate the nuclei from the cells. From these painstaking experiments, he identified a non-proteinaeceous compound that contained carbon, hydrogen, oxygen, nitrogen and a high amount of phosphorous. He named this substance "nuclein" because of its enrichment in nuclei and published a paper on his results in 1871 (Miescher 1871). In 1872, Miescher moved to the post of Professor of Physiology at the University of Basel and continued his work on nuclein. He found sperm from the salmon migrating in the Rhine or from the local salmon industry to be a convenient source to prepare nuclein in larger quantities and higher purity (Fig. 4.3, right-hand panel). Although Miescher suggested nuclein as the material of inheritance, he did not see how this single type of molecule could offer sufficient variety to allow for the diversity of species and in later years his research turned in other directions (Dahm 2005).

For many years after Miescher's discoveries, most researchers continued to consider that proteins must be the molecules of inheritance, because of their varied chemical structures based on twenty amino acids (note that twenty-two amino acids are known today). Nuclein (renamed as 'nucleic acid' in 1899) remained something of a curiosity until the 1940s and 1950s, when elegant experiments proved DNA to be the material of inheritance. Oswald Avery and colleagues established that bacterial transformation, (a process in which a property of one bacterial strain is conveyed to another strain in an inheritable manner), involved transfer of DNA (Avery et al. 1944). However, the experiment did not prove conclusively that DNA was responsible for the inheritable property. In the 1950s, Alfred Hershey and Martha Chase (Hershey and Chase 1952) carried out experiments in which they studied bacteriophages, viruses that infect bacteria, using biochemical methods of radioisotope labelling and cell fractionation. Bacteriophages were known to consist of DNA and protein, and Hershey and Chase "marked" the proteins and DNA of the bacteriophages by propagating them in media containing either the radioactive isotope of sulphur, ^{35}S, which would be incorporated into proteins, or the radioactive phosphorus, ^{32}P, which would be incorporated into DNA. By mixing bacteria with one or other set of the radio-labelled bacteriophages, allowing time for the 'phages to attach to the bacterial cells, and then fractionating the samples, they proved that it was proteins that served to attach the bateriophages to the outside of bacterial cells. The viral molecule that infects bacterial cells is DNA (Hershey and Chase 1952).

The next outstanding breakthrough was in 1953, in the form of a model for the structure of DNA. This model showed DNA to consist of two strands of covalently linked units (called deoxyribonucleotides) that wrap around each other in a double helix (discussed in more detail in the next section). The biochemists Linus Pauling and Francis Crick had originally proposed models for DNA structure that had the nucleotide bases pointing outwards. However, these did not accord with the excellent and careful unpublished X-ray data of the chemist and physicist, Rosalind Franklin (Fig. 4.4) and her student, Raymond Gosling, obtained while working with Maurice Wilkins at King's College London. Only the unorthodox ideas of Rosalind Franklin herself and James Watson (Fig. 4.5), an American geneticist working at that time with Francis Crick in Cambridge, that the bases are inside between the two helical strands, provided the correct answer.

The structure of DNA was of tremendous importance for solving many problems in biology at the level of how cells function. It also opened up the field of molecular biology. James Watson wrote a popular science book about the route to his model, *The Double Helix* (1968). This book also provided a view on the social relations between scientists including their strong competition and the prevalence of sexism. Franklin was not an author on the key publication in 1953, although, later, after

Fig. 4.4 Rosalind Franklin about 1950 X-ray diffraction of DNA

Fig. 4.5 James Watson in 1992 Model for the structure of DNA (Both from Wikipedia, English version 2020)

her death Francis Crick said that her contribution had been crucial (Maddox 2002). After moving to Birkbeck College, Franklin went on to study the crystal structure of tobacco mosaic virus and other plant viruses. In 1958, she died tragically of ovarian cancer at age 37. Well known is her question as a girl to her mother about god: How can I be sure that he is not she? Both scientists were atheists. Watson and Crick, with Maurice Wilkins, were honoured in 1962 by the Nobel prize in Physiology or Medicine, "for their discoveries concerning the molecular structure of nucleic acids and its significance for information transfer in living material."

4.3 DNA: The Molecule of Inheritance

Despite its simple chemical composition, the structure of DNA provides the basis for inheritance of traits, genetic variation between individuals of a species and genetic differences between species. DNA is made up of deoxyribonucleotides. Each nucleotide contains a deoxyribose sugar, phosphate and a base. The sugar-phosphate components are invariant and there are four different bases: adenine (A), thymine (T), guanine (G) and cytosine (C). In the structure of DNA, two lengthy, anti-parallel outer strands are formed by covalent bonding of the sugar group of one nucleotide to the phosphate group of the adjacent nucleotide; these strands twist around each other to form a helix (Fig. 4.6). The bases of each strand point into the central region of the helix and are linked to bases on the opposite strand by hydrogen bonds specifically to form the base pairs A-T, C-G, T-A or G-C (Fig. 4.7). Other stabilizing interactions within the double helix are the stacking interactions between the aromatic rings of the bases, which are arranged almost perpendicular to the helix axis (Figs. 4.6 and 4.7).

An extraordinary property of DNA is the process of DNA replication. With the help of enzymes, the two strands can be partially separated and untwined, and the complementary strands copied exactly, a mechanism of semi-conservative replication. The result is the synthesis of a new DNA molecule with the same sequence

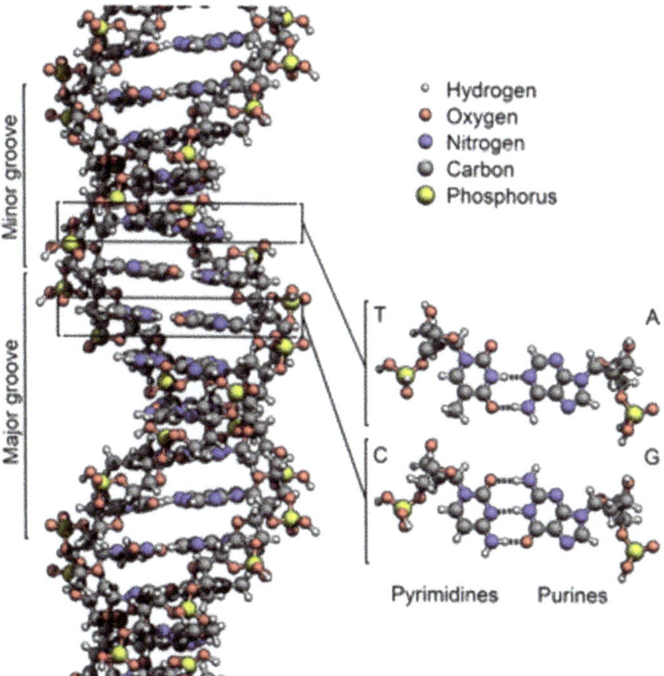

Fig. 4.6 Structure of the DNA double helix (From Wikipedia, English version 2020)

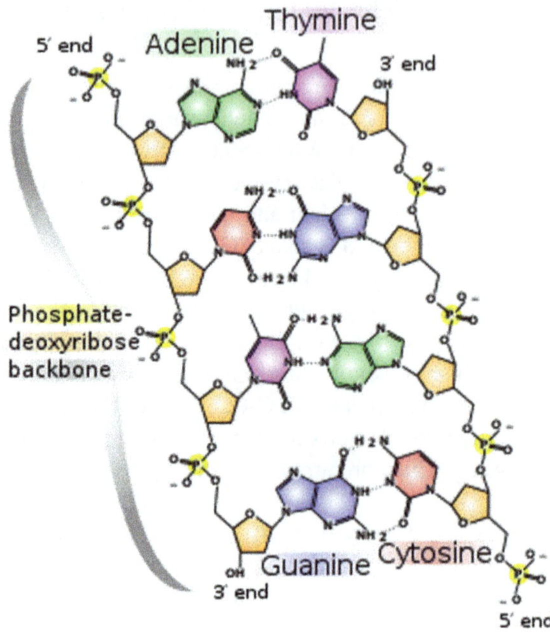

Fig. 4.7 Chemistry of the DNA double helix with longitudinal phosphate-deoxyribose backbone and complementary A-T, C-G, T-A or G-C base pairs (From Wikipedia, English version 2020)

of nucleotide bases. This process is only possible with a very low rate of replication errors. As discussed below, errors in replication can provide raw material for evolution.

In prokaryotic cells, the DNA forms a circular structure in the cytoplasm. In eukaryotes, DNA is located in the nucleus of a cell, tightly packed with proteins and protected in chromosomes. The human nuclear genome contains over 3 billion base pairs and would be ~ 2 metres long if laid out. It is one of the remarkable aspects of cell biology that this length of DNA exists compacted into the tiny nuclear volume of around 150 cubic microns. About 1% of the total genome is encoded by the much smaller mitochondrial genome that forms a closed circle and consists of about 16.5 kbase pairs. For protein synthesis, regions of the genome corresponding to "coding genes" are transcribed to messenger RNA (mRNA), which is transported to ribosomes in the cytoplasm at which translation to polypeptide chains (amino acids connected by peptide bonds) takes place with the help of transfer RNAs. In prokaryotes, which lack a nucleus, the DNA and RNA molecules are both contained in the cytosol. In all organisms including viruses, three bases form a triplet "codon" that determines which amino acid will be the result of translation. The list of triplets that specify the amino acids is called the triplet code. It is universal for all living organisms and is shown in Fig. 4.8. Note that some redundancy exists so that several triplets code for the same amino acid. Different organism may have different preferences for using a particular codon, a phenomenon referred to as codon bias. Remarkably, the relative availability of carbon or nitrogen in the local environment has been an evolutionary driver of the usage of A + T versus G + C bases in genomes, and is also a factor

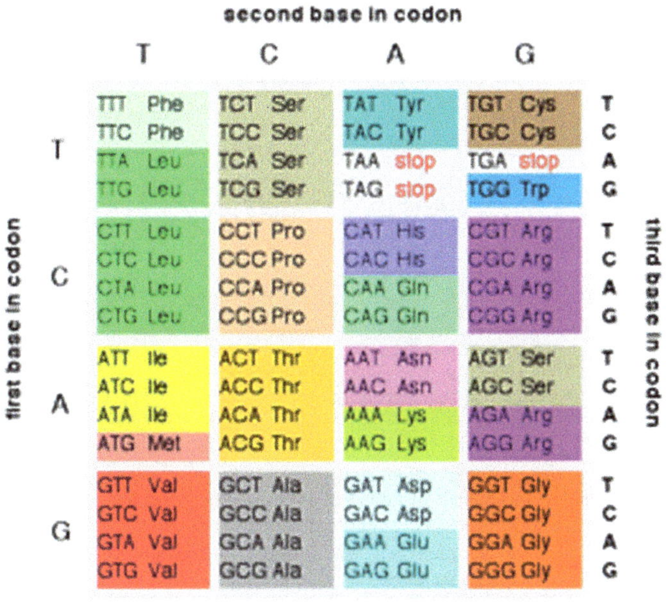

second base in codon

	T	C	A	G	
T	TTT Phe	TCT Ser	TAT Tyr	TGT Cys	T
	TTC Phe	TCC Ser	TAC Tyr	TGC Cys	C
	TTA Leu	TCA Ser	TAA stop	TGA stop	A
	TTG Leu	TCG Ser	TAG stop	TGG Trp	G
C	CTT Leu	CCT Pro	CAT His	CGT Arg	T
	CTC Leu	CCC Pro	CAC His	CGC Arg	C
	CTA Leu	CCA Pro	CAA Gln	CGA Arg	A
	CTG Leu	CCG Pro	CAG Gln	CGG Arg	G
A	ATT Ile	ACT Thr	AAT Asn	AGT Ser	T
	ATC Ile	ACC Thr	AAC Asn	AGC Ser	C
	ATA Ile	ACA Thr	AAA Lys	AGA Arg	A
	ATG Met	ACG Thr	AAG Lys	AGG Arg	G
G	GTT Val	GCT Ala	GAT Asp	GGT Gly	T
	GTC Val	GCC Ala	GAC Asp	GGC Gly	C
	GTA Val	GCA Ala	GAA Glu	GGA Gly	A
	GTG Val	GCG Ala	GAG Glu	GGG Gly	G

first base in codon — third base in codon

Fig. 4.8 The universal genetic code

in the evolution of codon usage (Shenhav and Zeevi 2020). Most importantly, the universality of DNA and the triplet genetic code in all living organisms provides a direct proof of Darwin's second postulate that all modern organisms descended from a common ancestor.

The complete DNA content of an organism is referred to as its genome and the different transcribed regions are termed genes. Another important feature that has been explored by molecular genetics (a branch of molecular biology), is the nature of mutations, i.e. changes to the sequence of bases. Many mutations involve deletion or insertion of a region of DNA, which would profoundly affect the function of any encoded protein. More minor mutations, termed point mutations, change a single base and yet are also prominent causes of inherited diseases. If a single nucleotide base is altered within a protein-encoding gene (Fig. 4.9), it may result in either no change to the protein (because of the redundancy of the genetic code (Fig. 4.9), the coding of a different amino acid at that position in a protein, or may result in a "STOP" codon, which terminates translation of the polypeptide at that position.

Point mutations can exist within the genomes of the individuals that make up the population of a species without pathological consequences. In this instance, the altered sequence is generally referred to as a genetic polymorphism and is one of the sources of the genetic variation that allows for evolution. In the case of inherited diseases of humans, some diseases have been identified to be caused by a point mutation in one particular gene. One example is cystic fibrosis, a genetic disorder in which excess mucus builds up in the lungs, leading to lung infections and breathing

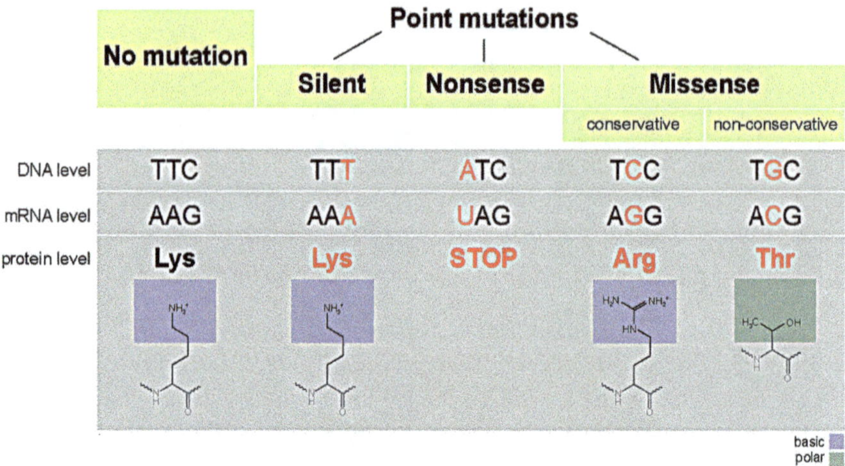

Fig. 4.9 How point mutations lead to amino changes in a polypeptide sequence. Amino acid substitutions may be conservative (ie, an amino acid with the same chemical properties as the wild-type) or non-conservative (ie, an amino acid with different chemical properties to the wild-type). (From Wikipedia, English version 2020)

difficulties. This pathology is caused by mutations in the gene that encodes the cystic fibrosis transmembrane conductance regulator protein (CFTR). This protein forms a channel in the plasma membrane of lung epithelial cells and regulates the passage of chloride ions in and out of the cells. If the channel does not function correctly, water balance in the airways is affected leading to sticky mucus. Several types of mutations cause cystic fibrosis with different severities of disease (Fig. 4.10). Most of these are point mutations that substitute amino acid an within the coding sequence of CFTR, leading to different effects on the function of the channel. Deletion of Phe508 (F508del in Fig. 4.10) has very severe consequences because this mutation affects the folding of CFTR protein so that it is not transported to the plasma membrane and so is completely non-functional (Fig. 4.10) (Marunaka 2017).

At a first glance, mutations have disturbing effects. However, by giving rise to genetic variations between individuals or groups of a species, they provide a source for the genetic diversity on which natural selection acts. Many mutations have no apparent consequence, but a mutation might lead to an alteration in gene expression or protein function which gives individuals carrying the mutation an improved adaptation to their environment, and thus an increased likelihood of surviving and passing the mutation on to their offspring. Mutations, and thus, in general terms, changes or errors in the transfer of information, are therefore necessary for evolution.

How do mutations arise? Mutations can be caused by a number of external events, of which high-energy radiation by UV-light, X-ray or γ-radiation, are the most significant. A natural example is the exposure to UV in sunlight. Such forms of radiation also emerge at high intensities from the radioactive waste of atomic bombs or nuclear fuels, or radioactive pollution of large areas by military tests. Particularly harmful

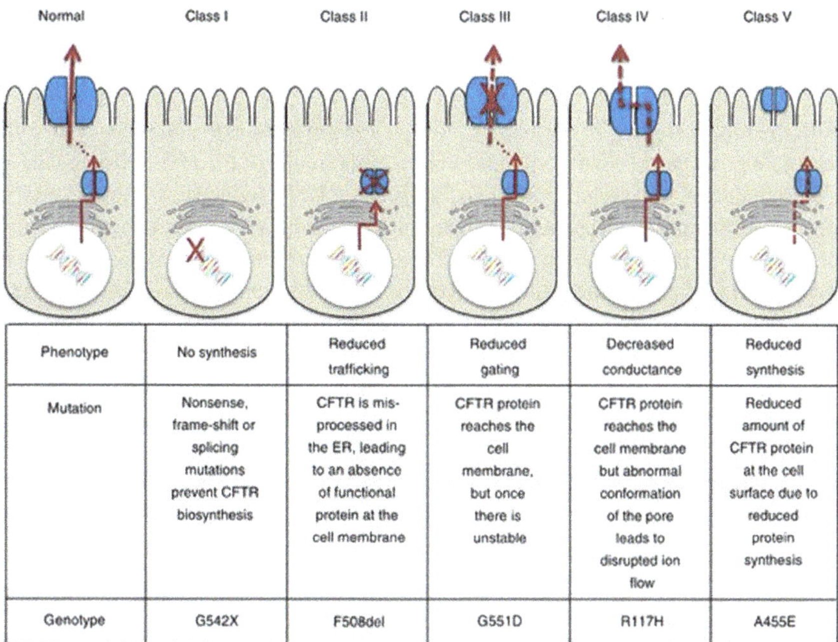

Phenotype	No synthesis	Reduced trafficking	Reduced gating	Decreased conductance	Reduced synthesis
Mutation	Nonsense, frame-shift or splicing mutations prevent CFTR biosynthesis	CFTR is mis-processed in the ER, leading to an absence of functional protein at the cell membrane	CFTR protein reaches the cell membrane, but once there is unstable	CFTR protein reaches the cell membrane but abnormal conformation of the pore leads to disrupted ion flow	Reduced amount of CFTR protein at the cell surface due to reduced protein synthesis
Genotype	G542X	F508del	G551D	R117H	A455E

Fig. 4.10 Effects of point mutations on the synthesis and function of CFTR. Reproduced from Koivula, F.N.; McClenaghan, N.H.; Harper, A.G.; Kelly, C. Islet-intrinsic effects of CFTR mutation. Diabetologia 2016, 59, 1350–1355. Open access under the terms of the Creative Commons Attribution 4.0 International License

are radioactive compounds such as barium carbonate, which deposits in the bones of vertebrate organisms. Radioactive barium has a very long half-life and can be lethal due to its long lasting mutational activity. Chemicals can also cause mutations, and can affect nucleotide base pairing or DNA replication by intercalating between nucleotide bases. These mutagens can potentially be taken up from the environment or in food. An example found in research laboratories is ethidium bromide, a widely used laboratory chemical that intercalates between nucleotide bases, resulting in insertion or deletion of a base pair when the DNA is next replicated. Mutations can also arise due to biological errors in DNA replication or transcription. Although DNA replication is a very accurate process, rare errors occur at a frequency of around one incorrect nucleotide per 10^8–10^{10} nucleotides polymerised (Eigen 1992). The human germline mutation rate is approximately 0.5×10^{-9} per basepair per year and differs significantly from the rate estimated for other animals and plants (Scally 2016). The general high accuracy of DNA replication, transcription, and translation to proteins is achieved by high accuracy of the DNA polymerases and sophisticated, robust "proof-reading" and correction mechanisms that operate at each stage of gene expression.

4.4   Pathways of Evolution Modelled from Phylogenetic Plots: Advantages and Unproven Assumptions

With the great progress of molecular biology and genetics in the last forty years, the complete genomes of many organisms from all domains of the tree of life have now been sequenced. Through these efforts, millions of DNA sequences for genes coding for related or unrelated proteins in different species have became known. Because the triplet code is universal, the amino acid sequences of the encoded proteins can be predicted from the genome sequence. Remembering that evolution occurs at the level of the genes and the encoded proteins, this mass of DNA sequence information gives us a view on the pathways of evolution, the relationships between modern species, and a means to estimate the timing of the evolution of different genes. Assuming that homologous proteins changed with a fixed mutation rate during evolution, differences between sequences in different species can be taken as proportional to the time since the divergence of one protein from other. Differences between the sequences of related proteins therefore contain information on phylogenetic differences. The sequences can provide a kind of clock (termed the molecular clock), which is a record of the timing of evolution between the organisms that are alive today.

The problem is how to extract this information? A critical evaluation is necessary because many biologists tend to use widely available computer programs and the assumptions that are part of the programs need to be considered. It is a property of all phylogenetic programs that the output will provide a phylogenetic tree, even if the supporting evidence is very weak or incorrect assumptions are included.

The accuracy of phylogenetic trees depends on the homology of the proteins to be compared: it is useless to calculate phylogenetic distances between sets of unrelated proteins. The sequences must have sufficient overall similarity to prove that the proteins have a common origin. A given amino acid residue should correspond to the same residue in the compared protein. This is not always easy to verify and complex computer evaluations have been developed to obtain the best sequence alignment with matching of homologous regions. Particularly difficult is the handling of long proteins with homologous regions that are interrupted by loops without homology. One example is the method of Landan and Grauer (2007), used in the LUCA hypothesis introduced in Chap. 3. A sequence identity greater than 20% is often set as a threshold for determining if homology exists. However, neither this estimate nor the method of Landa provides absolute certainty of homology. From our own experiences, error in the alignment of homologous regions is a major reason for incorrect results. Independent evidence, such as the existence of an experimentally determined structure for a protein domain or set of domains, can bring much greater confidence to assessments of domain homologies and the boundaries of domains.

Assuming that the compared regions are indeed homologous, a phylogenetic tree can be constructed. To the extent possible, the compared sequences should include proteins from organisms that represent different branches of evolution in order to achieve a reasonable error rate. The ratio between the sequence difference and the rate of mutations is the time distance between two proteins. These distances are plotted in two dimensions, minimizing the total distance between all proteins.

Remarkably, a speculative phylogenetic tree was constructed by Charles Darwin (1859), in this case based on morphological similarities between species. Trees based on morphological or anatomical characteristics of organisms are in broad agreement with trees based on DNA or protein sequence differences, but the advent of whole-genome sequencing has brought many changes and much greater precision to our understanding of the taxonomical relationships between species or phyla.

The literature contains examples of rooted and unrooted phylogenetic trees. Trees calculated only from sequence differences are unrooted trees. In this context, it is uncertain in which direction evolution took place, or in other words, these trees do not contain information to identify which species arose earlier on the Earth than others. To enable rooting, this information may be drawn from independent data such as the fossil record or known taxonomical relationships. For example, it is rather clear that humans evolved later than bacteria. In the case of a tree built from proteins present in both humans and bacteria, the tree should be rooted on bacteria. In most cases, the interest in reading a tree relates to the direction of increasing time of evolution. In these cases phylogenetic trees are a reliable tool which teaches us about the timing of evolution, at least in a global way. A much higher risk of misinterpretation exists for the reverse case, namely reading the trees from its leaves to the root. This is done when finding the hypothetical root by extrapolation. Examples for such extrapolations are the LUCA hypothesis and the Urgen of Eigen (Chap. 3). Typically, trees based on molecular data tend to place divergences between groups of organisms at earlier times than those derived from the fossil record. Figure 4.11 points up these difficulties in a tree showing the phylogenetic relationships of different clades and phyla of animals which is scaled to timelines derived either from known fossils or from sequence-based phylogenetic analyses with calculated molecular clocks (Cunningham et al. 2017). It can be seen that the earliest evolutionary divergences have the widest range of molecular clock dates.

A major problem in phylogeny is extrapolation into regions in which the necessary conditions no longer hold. In the case of extrapolations of phylogenetic plots, the assumption that the mutation rate has been constant over time is the critical parameter. Some phylogenetic programs introduce changes in rate for different bases and amino acids, but a fundamental problem is that we cannot know the mutation rate that existed at the time at which the hypothetical Urgen or LUCA (Chap. 3) evolved. One cause of mutations is radiation damage, yet atmospheric radiation at the Earth's surface is considered to have changed considerably over the evolutionary history of life. For other reasons, the reliability of back-extrapolation to early DNA or protein sequences decreases exponentially with the distance between the present state and the ancestral state. For example, in many cases it is not known whether all proteins selected for analysis evolved by the same pathway or were formed by distinct gene fusion or domain reshuffling events. If a gene duplicated in the remote past, it may be almost impossible to assign modern genes to one or other duplication product.

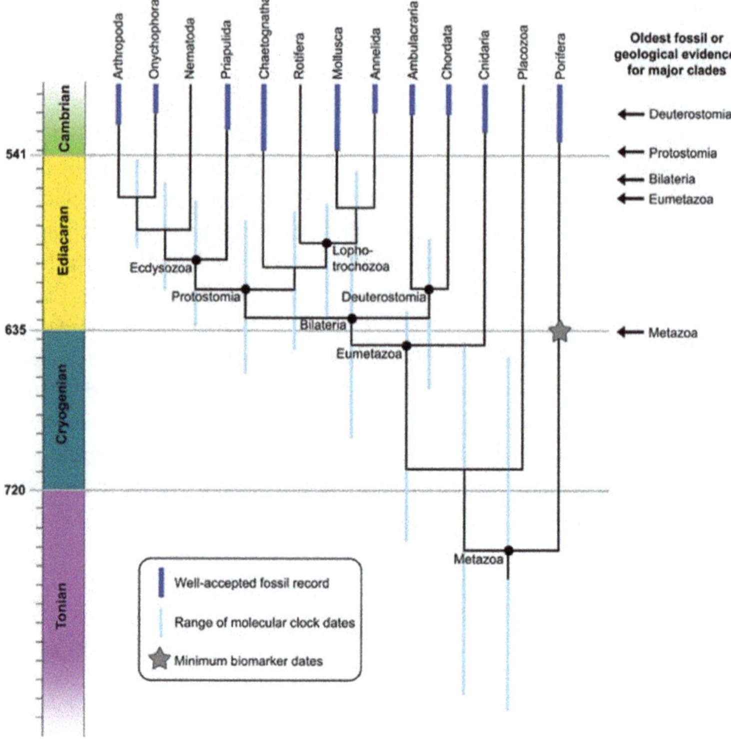

Fig. 4.11 Phylogenetic tree of animal evolution, scaled in millions of years in relation to fossil data and calculated molecular clocks. Reproduced from Cunningham et al. (2017), Bioessays 39, 1–12, under Attribution 4.0 International (CC BY 4.0)

4.5   The Human Genome Contains Genes from Different Ancestral Origins

The fascinating question of how humans evolved, which has traditionally relied on the fossil record, has also benefited greatly from molecular biology and genomics. Amongst the modern primates, humans are most closely related to chimpanzees. Using the phylogenetic methods discussed above, the human lineage is estimated to have diverged between 5 and 8 million years ago (Wood and Richmond 2000). The fossil record shows that many species of "hominins" (ancestral human-like primates) existed for millions of years before our species, *Homo sapiens,* emerged around 200,000 years ago in Africa. The oldest known fossils confidently assigned as hominins are of *Ardipithecus ramidus* from Ethiopia, dated to about 4.5 million years ago, and *Australopithecus anamensis* from Kenya, dating from 3.9 to 4.2 million years ago (Wood and Lonergan 2008). However, there are difficulties in deciding from fossil evidence alone whether the fossils are from hominins or from extinct relatives of great apes, and how many different species the fossils may represent. Several species of the genus *Homo* are also recognised, for example *Homo erectus*, who originated in Africa several million years ago and the more recent Neanderthals, *Homo neanderthalensis*, identified in Europe, who became extinct about 40,000 years ago. Human evolution is currently viewed not as a linear progression, but as several bursts of radiation during which inter-breeding took place between early hominins or, later, different species of the *Homo* genus (Fig. 4.12). It is suggested that global climatic cooling

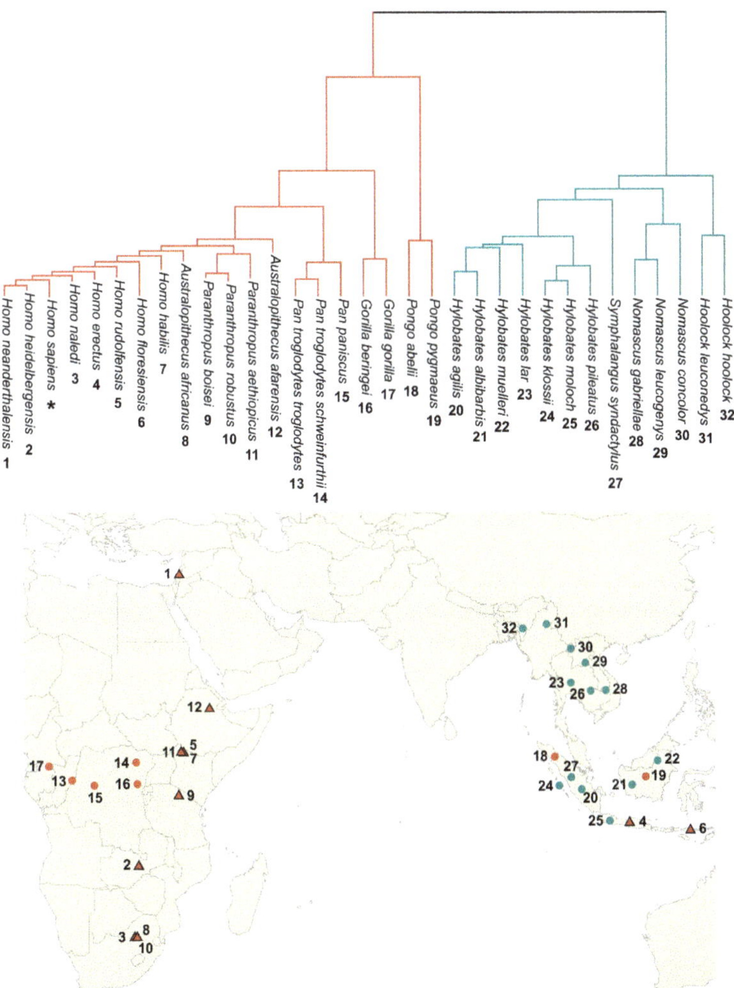

Fig. 4.12 Phylogenetic plot of human evolution based on craniofacial morphologies. Hominidae (including gorilla, chimpanzee and orangutan) are in red and Hylobatidae (gibbons) in blue. Numbers refer to the locations of specimens as shown on the world map (lower panel). On the map, fossils are shown as triangles and extant species as circles. Reproduced from Rocatti and Perez (2019), Scientific Reports 9:15,267 under open access Creative Commons Attribution 4.0 International Licence

could have been an important factor in driving some species of *Homo* to extinction (Raia et al. 2020).

The ability to extract ancient DNA from bones or teeth and to obtain high-throughout DNA sequence data from tiny DNA samples has revolutionised paleontological research. Draft genome sequences have been obtained from bones from Neanderthal burial sites (around 38,000 years old) and from the Denisovans, a much earlier group of hominins first identified in Siberia (existing about 50,000–100, 000 years ago; reviewed by Llamas et al. 2017). In 2020, Denisovan DNA was also recovered from materials in Baishiya Karst Cave on the Tibetan Plateau, with indications that they had used the cave for shelter over centuries between 100,000 and 45, 000 years ago (Zhang et al. 2020). Remarkably, it has also been possible to sequence nuclear DNA from 400,000 year old hominin remains from Spain (Meyer et al. 2016). These molecular studies give us new insights into how these groups may be related to each other and to modern humans, and also about places where *Homo* species interacted with each other.

It is well documented in the fossil record that humans evolved in Africa and migrated at various times to different parts of world. The sequencing of many genomes from modern people in different world regions has given insight into the geographical origins of genetic polymorphisms in the global human population. This has set the scene for an interesting field of genetics that involves tracing back the origins of human genes. Today, an individual can order DNA sequencing of a part of their genetic information from a commercial company, (as one example, the company 23andme https://www.23andme.com/en-int/), to gain an insight into whether some of their genes include variants associated with common diseases. Considering the sequencing of the whole of a person's genome, a typical North American might find out, for example, that 30% of their genes came from Europe, where their family lived for a long time, 10% from native Americans, and 60% from other regions of the world. These numbers are given only as a demonstration and do not belong to a real person. The concept of genetically distinct "races" is completely ruled out by these results. Instead human genomes are colourful mixtures of very many sources and a shared ancestry.

Along the same lines, the Swedish geneticist Svante Pääbo (Fig. 4.13) has investigated what fraction of the genome of Neanderthals is still present in an average European. There is evidence that Neanderthals mated with modern humans (*H. sapiens*) who had migrated from Africa into Europe. Pääbo and others were able to show that 2% to 3% of the sequence of the Neanderthal genome is preserved in the average contemporary European genome. All non-African people also carry some DNA sequences (less than 1% of the whole genome) from at least two sources in the remote Denisovans. People from Papua carry up to 6% Denisovan DNA inherited from one of these sources (reviewed by Llamas et al. 2017). Pääbo and colleagues also identified DNA from a 34,000 skull fragment discovered in Mongolia, to include Denisovan DNA segments that are related to those found in modern Asians (Massilani et al. 2020). All these results give us profound insights into how early humans mingled over extended timeperiods and spread into Europe or Asia. A book by Pääbo

Fig. 4.13 Svante Pääbo. Co-director of the Max Planck Institute for Evolutionary Anthropology in Leipzig, Germany (From Wikipedia, English version 2020)

(2014) is a bestseller and contains many interesting ideas. He is probably one of the most frequently cited geneticists of our time.

With the technological improvements and reduced costs of genome sequencing, it has become possible for increasing numbers of modern human genomes to be compared in the same study. A recent analysis of 929 genomes sequenced from 54 geographically and culturally diverse human populations has revealed vast amounts of individual genetic variation in the form of single-nucleotide polymorphisms and provides more evidence for a complex history of *H. sapiens* mixing with the Denisovans (Bergstrom et al. 2020).

References

Avery O, MacLeod C, McCarty M (1944) Studies on the chemical nature of the substance inducing transformation of pneumococcal types. J Exp Med 79:137–158

Barlow N (ed) (1958) The autobiography of Charles Darwin 1809–1882. W. W. Norton and Co, New York

Bar-On YM, Phillips R, Milo R (2018) The biomass distribution on Earth. Proc Natl Acad Sci U S A 115:6506–6511. https://doi.org/10.1073/pnas.1711842115

Bateson W (1901) Problems of heredity as a subject for horticultural investigation. J R Horticu Soc 25:54–61

Bergström A, McCarthy SA, Hui R, Almarri MA, Ayub Q, Danecek P, Chen Y, Felkel S, Hallast P, Kamm J, Blanché H, Deleuze JF, Cann H, Mallick S, Reich D, Sandhu MS, Skoglund P, Scally A, Xue Y, Durbin R, Tyler-Smith C (2020) Insights into human genetic variation and population history from 929 diverse genomes. Sci 20;367(6484):eaay5012. https://doi.org/10.1126/science.aay5012

Cunningham JA, Liu AG, Philip SB, Donoghue CJ (2017) The origin of animals: can molecular clocks and the fossil record be reconciled? BioEssays 39:1–12

Dahm R (2005) Friedrich Miescher and the discovery of DNA. Dev Biol 278:274–288. https://doi.org/10.1016/j.ydbio.2004.11.028

Darwin C (1839) Voyage of the Beagle. Penguin Books, 1989

Darwin C, Wallace A (1858) On the tendency of species to form varieties; and on the perpetuation of varieties and species by natural means of selection. Zool J Linnean Soc 3(9):45–62

Darwin C (1859) The origin of species, New edition 1979 Gramercy, Random House, New York

Eigen M (1992) Steps towards life. Oxford University Press, 1–173

Hershey AD, Chase M (1952) Independent functions of viral protein and nucleic acid in growth of Bacteriophage. J Gen Physiol 36:39–56

Koivula FNM, McClenaghan NH, Harper AGS et al (2016) Islet-intrinsic effects of CFTR mutation. Diabetologia 59:1350–1355. https://doi.org/10.1007/s00125-016-3936-1

Landan G, Grauer D (2007) Heads or tails: a simple reliability check for multiple sequence alignment. Mol Biol Evolution 24:1380–1383

Llamas B, Willerslev E, Orlando L (2017) Human evolution: a tale from ancient genomes. Philos Trans R Soc Lond B Biol Sci. 372(1713):20150484. https://doi.org/10.1098/rstb.2015.0484

Maddox, B (2002) Rosalind Franklin: the dark lady of DNA. HarperCollins

Marunaka Y (2017). The mechanistic links between insulin and Cystic Fibrosis Transmembrane Conductance Regulator (CFTR). Cl—Channel Int J Mol Sci 18(8):1767. https://doi.org/10.3390/ijms18081767

Massilani D et al (2020) Denisovan ancestry and population history of early East Asians. Sci 370:579–583. https://doi.org/10.1126/science.abc1166

Mayer E (1997) This is biology, the science of the living word. Harvard University Press

Meyer M, Arsuaga J-L, de Filippo C, Nagel S, Aximu-Petri A, Nickel B, Martínez I, Gracia A, Bermúdez JM, de Castro E, Carbonell BV, Kelso J, Prüfer K, Pääbo S (2016) Nuclear DNA sequences from the Middle Pleistocene Sima de los Huesos hominins. Nat 531:504–507

Miescher F (1871) Ueber die chemische Zusammensetzung der Eiterzellen. Med Chem Unters 4(1871):441–460

Pääbo S (2014) Neanderthal man: in search of lost genomes. Basic Books

Raia, P, Mondanaro, A, Melchionna, M, Di Febbraro, M, Diniz-Filho, JAF, Rangel, TF, Holden, Philip, Carotenuto, F, Edwards, Neil, Lima-Ribeiro, MS, Profico, A, Maiorano, L, Castiglione, S, Serio, C and Rook, L (2020) Past extinctions of Homo species coincided with increased vulnerability to climatic change. One Earth 3(Early access). https://doi.org/10.1016/j.oneear.2020.09.007

Scally A (2016) The mutation rate in human evolution and demographic inference. Curr Opin Genet Dev 41:36–43

Shenhav L, Zeevi D (2020) Resource conservation manifests in the genetic code. Sci 370:683–687

Wallace, AR (1855) On the law which has regulated the introduction of new species. Ann Mag Nat Hist (ser.2) 16(93):184–196

Watson JD, Crick CH (1953) A structure for deoxyribonucleic acid. Nat 171:371–378

Watson JD (1968) The Double Helix: a personal account of the discovery of the structure of DNA. Athenium, New York
Wood B, Richmond BG (2000) Human evolution: taxonomy and paleobiology. J Anat 197:19–60
Wood B, Lonergan N (2008) The hominin fossil record: taxa, grades and clades. J Anat 212(4):354–76. https://doi.org/10.1111/j.1469-7580.2008.00871.x.
Zhang D et al (2020) Denisovan DNA in Late Pleistocene sediments from Baishiya Karst Cave on the Tibetan Plateau. Sci 370:584–586. https://doi.org/10.1126/science.abb6320

Chapter 5
Manipulated Evolution and Artificial Life

Since the 1970s, laboratory research methods have been developed that allow for the manipulation of the genes or genome of an organism. These techniques, which go under the broad name of molecular biology, allow scientists to segregate, delete, or modify specific regions of DNA, in particular parts of genes. These methods can be applied to cells cultured in a laboratory, or to whole organisms under controlled, laboratory conditions. The latter approach has been applied to generate genetic clonal variants of many organisms, principally bacteria, fungi, or certain animals or plants. The genetic modification of an intact organism means that the genetic change would be propagated to their offspring. From an ethical standpoint, such manipulations have different levels of risk. Genetic changes made to a single organism in a laboratory, or a culture of cells, are not propagated to wider populations as long as the modified organism(s) remain securely in the laboratory. In contrast, inheritable changes in the genome of an organism that could enter and inter-breed with natural populations have the potential to affect biological evolution by natural selection.

The risks connected with genetic engineering that leads to new variants of a species are manifold, and despite legal frameworks, unethical, military or criminal uses are of concern. The engineered organisms have not evolved under natural selection as a result of a close connection with their natural environment. If they are released from the laboratory (either accidently or by design), their interaction with the complex, natural ecological networks is unpredictable. In most countries, genetic engineering processes are under strict legal control, which covers commercial, agricultural, or disease control applications. However, there are some cases of approved deliberate release of genetically engineered organisms into the wild for a specific purpose. An example is the release of genetically engineered *A. aegypti* mosquitoes to control the spread of dengue fever, an increasingly prevalent disease caused by the dengue virus. The engineered male mosquitos mate normally with female mosquitos in the wild, but their offspring die before adulthood due to insertion of a "suicide gene" in the genome of the males. This trial was carried out in a suburb of Juazeiro, Brazil and caused the wild mosquito population to decrease over several months (Carvalho et al. 2012). The strategy of release of large numbers of these genetically modified mosquitos was

J. C. Adams and J. Engel, *Life and Its Future*,
https://doi.org/10.1007/978-3-030-59075-8_5

discussed and agreed in advance with the local community. Another approach that is being tested (although not yet at the stage of a controlled trial), is that of the "gene-drive". In this method, relatively small numbers of *A. aegypti* mosquitoes engineered to express a molecule derived from an antibody that binds to dengue virus would be released. Through mating, the genetically engineered mosquitos would transmit the gene throughout wild populations of *A. aegypti*. The expressed antibody-like molecule would act to block viral proliferation or movement of viral particles within the body of virally-infected mosquitoes (termed "neutralisation" of the virus), and so reduce the number of viral particles in mosquito saliva. The concept is that the transmission of dengue virus from mosquitos to humans would be reduced (Buchman et al. 2020).

A second important risk emerges from possible errors in the engineering methods themselves. Errors could lead to additional, unexpected changes in a genome which could have unknown or harmful effects. However, as the costs of DNA sequencing have decreased over time, it has become routine to sequence an entire genome, which makes it easier to monitor for unwanted genetic changes and to work with cells or organisms that include only the designed genetic change. Another possible risk is that, if a genetically-modified organism is released outside laboratories (either accidently or for a presumed beneficial purpose), it could have an unexpected biological advantage or higher reproductive rate than the natural species. Either of these advantages would enable the engineered organisms to compete against the natural, wild-type members of the same species in an uncontrolled way. A relevant example (although not a deliberately engineered organism) is provided by *Procambarus virginali*, a new species of marbled crayfish, that emerged in Germany from the aquarium trade in 1995. Very unusually, this crayfish has a triplod genome (i.e., three copies of each gene; animals are in general diploid with two copies of each gene), and has gained the ability to reproduce by parthenogenesis (the development of an embryo from an unfertilised egg cell). Through the pet shop trade and releases to local environments, this crayfish has spread internationally very rapidly and has established large populations in the wild. Individual marbled crayfish from different countries have indistinguishable genomic DNA sequences, providing evidence that this new species has invaded rapidly around the world as a clonal organism, through multiplication by parthenogenesis (Gutekunst et al. 2018). The new crayfish species provides a view on how the release of a genetically modified organism(s) to the environment could have far-reaching and unexpected consequences (Fig. 5.1).

Unfortunately, confusing and often non-relevant discussions about the issues raised by genetic engineering occur frequently in public media and politics and difficulties are often compounded by a lack of scientific knowledge. In the following sections we shall briefly review examples of commonly used methods of genetic engineering. The present text does not attempt a full discussion of genetic engineering methods or the ongoing analysis by ethical commissions. It may, however, help to clarify some basic issues. A nice book that describes new activities that may change the world has been written by Christopher Preston (2018) and are also discussed in his article of 2019.

Fig. 5.1 The marbled crayfish, *Procambarus virginalis*, in a laboratory setting. This new species has spread rapidly around the world, probably due to its triploid genome and ability to reproduce asexually. Scale bar = 1 cm. (Reproduced from Gutekunst et al. [2018], under open access)

5.1 Expression of Recombinant Proteins in Bacteria

This method, often called recombinant DNA technology, opened up very many research directions in molecular biology about 30 to 40 years ago. JE remembers this breakthrough period in his own research very clearly. JE was active through his research career in the field of the extracellular matrix of animals (Engel 1992, 2017). The extracellular matrix contains many large and complicated proteins that are deposited between cells and influence cell function in many ways. Most of these proteins consist of many domains (ie., self-folding regions of the protein) which have different functions. The matrix proteins are therefore called multifunctional proteins. In the beginning of JE's investigations it turned out to be very difficult, if not impossible, to isolate these proteins from animal tissues by biochemical methods. Furthermore, no powerful methods existed to study the function of the individual domains. For JE's work and for the work of many other biologists, it was a tremendous advantage to gain the ability to express either the full-length protein or the component domains by use of recombinant DNA technology in bacterial systems.

In brief, to achieve expression of a protein from a bacterial plasmid, a DNA or RNA fragment needs to be prepared from a cell type that contains the genetic information for the protein or domain of interest. With a number of experimental tricks in the laboratory, this fragment can be inserted into plasmid DNA from a suitable (not pathogenic), easily grown bacterium. The species *Escherichia coli*, that is part of the normal gut microbiota of humans and some other animals, has become the most-used laboratory model for these studies (Blount 2015). The *E. coli* bacterial culture is then expanded so that, as the bacteria grown and divide, they produce the encoded protein in large amounts together with their own proteins. Following suitable biochemical and biophysical isolation procedures, it is possible to produce and purify 100 mg or more of the desired protein from a few litres of bacterial culture. This is enough material for a large number of functional studies. With the pure protein, it also becomes possible to elucidate the three-dimensional structure of the proteins or its domains by X-ray crystallography, a method which requires a large number of sample molecules.

The method of recombinant DNA technology now exists in very many variant forms. For example, artificial mutations can be introduced into protein-coding sequences. This is useful for the identification of the active sites of enzymes or protein–protein binding sites. Encoded amino acid residues can either be changed to other residues or completely deleted (see also Chap. 4, Sect. 4.3). Any consequent effects on activity or properties of the protein may be monitored by laboratory experiments. Also, insoluble proteins may be rendered soluble for purification by fusing the DNA coding sequence with DNA encoding a hydrophilic C-terminal sequence.

Much ethical debate surrounded the emergence of these molecular biology technologies. The Asilomar conference of 1975 was called by scientists themselves after a self-imposed worldwide moratorium on the area of research, and was a watershed for the whole field. The conferees decided that genetic engineering experiments could proceed, but under very specific guidelines (Berg 2008). However, for all scientists who obtained exciting new results with the novel technology and could see the potential for beneficial applications, it was a shock when politicians tried to ban genetic methods on a wide scale based on ethical arguments. J.E. remembers the initiative in his home country Switzerland "for the protection of life and neighbourhood against gene manipulations". The legislation asked applicants for a proof of usefulness and safety, and a lack of alternatives, as well as an outline of ethical accountability. The following text (here translated to English; the official versions were in German and the three other languages of Switzerland) was suggested as a public vote to the Swiss nation in October 1998:

"Federal initiative for the protection of life and environment by gene manipulations (gene protection initiative).

To the federal constitution a new article (Art. 24^{decies}, new) will be added: The federal government releases instructions against misuses and dangers caused by genetic alterations of the genomes of animals, plants and other organisms. Thereby the dignity and meaning of living species, the maintenance and use of genetic diversity as well of the safety of humans, animals and environment should be considered.

Forbidden are:

1. Manufacturing, acquisition and transfer of genetically altered animals
2. Release of genetically altered animals to the environment
3. Granting of patents for genetically modified animals and plants and their constituents as well as for applied procedures and products.

The Legislation contains directions about:

1. Manufacturing, acquisition and transfer of genetically modified plants
2. Industrial production of compounds and applications of genetically modified organisms
3. Research with genetically altered organisms, which constituent a risk for human health and environment."

In October 1998, many thousands of biologists (including JE) demonstrated in Zürich to argue that including this text in the constitution of Switzerland would block biological science research in Switzerland completely. In fact, the gene protection initiative was rejected by the Swiss people, as only 33% of them supported the initiative. Looking back on this time, the dangers of genetic engineering were overestimated. People have learned to discriminate between aspects of gene manipulations that are dangerous or unpredictable and those that can safely be carried out in the laboratory, and the legislation has been adapted to this. For example, plasmid-based methods to express proteins in bacteria as performed under the control of good laboratory practice qualifies as a potentially harmless method, providing that the protein studied does not have acute human health implications.

In Switzerland, all manipulations with human genes that result in hereditable changes are forbidden or strongly controlled. Similar manipulations with plant and animal genes are also controlled but much less strictly. We shall discuss these aspects further in Sects. 5.3 and 5.4. A major problem internationally is the very different, or even non-existent, legislation of genetic engineering in different countries around the world. Given the globalisation of many commercial interests, this renders the prevention of misuse of genetic engineering rather inefficient.

5.2 Transgenic Mice and Gene Knockout Mice

The genetic engineering of mice, zebrafish, fruit flies, worms or other model organisms with artificially introduced genetic deletions of gene regions (often called gene knockouts) is a commonly used method in modern molecular biology. The main purpose is to identify the function of a genetic region or gene product in a living organism. A typical and frequent research problem is to identify the biological function of a protein of interest, for which the location and sequence of the encoding gene within the genome is known (given that complete genome sequences are now available for many organisms). To discover the function of a gene product in a mammalian species, generating a "gene knockout" mouse (i.e., a mouse in which a part of the genomic sequence encoding the protein of interest has been deleted, so that the protein cannot be synthesised by the cells of the mouse) can provide an answer. Several types of results have been observed in these studies:

1. The removal of the encoded protein leads to the death of the mouse at an embryonic stage. This proves that the natural protein has an essential role in embryonic development.
2. In comparison to the wild type mouse, the gene knockout mouse suffers from a severe disease condition, which in most cases is inherited to its offspring. A careful investigation of the organs involved and the cellular basis of the disease may then lead to detailed insights concerning the function of the protein, and possible new insights into related diseases in humans.

3. The gene knockout mouse appears to be unchanged from the wild type mouse. This does not mean that the studied gene product has no function. The lack of a phenotypic change may be explained by the presence of a second gene with similar function (for example, a related member of a gene family that can compensate), or that the effect of deleting the studied gene was subtle, or will be apparent only at later life stages, or under certain environmental conditions. It should be remembered that the laboratory mice live under conditions where their survival needs are met and they do not need to hunt for food or water. In these living conditions, the loss of function of a particular protein may be tolerated, whereas in the wild, the gene knockout might reduce the health and fitness of the mouse. This can be tested experimentally by placing the knockout mouse under conditions that challenge its fitness or agility, or under conditions that could reveal reduced health, such as exercise systems or a particular diet. Such types of experiments have indeed revealed important functions for many gene knockouts that at first glance had very limited phenotypic effects.

In scenarios 2 and 3, because the mutation has been introduced into the germ line, the offspring of the knockout mice will inherit the introduced gene deletion. These genetically altered mice must be kept in animal banks and not released outside the laboratory. This is difficult to achieve or monitor because many of these mouse lines are commercially available and studied in many laboratories all over the world. An overall global control is missing.

Similar manipulations have been performed with other vertebrate or invertebrate animals (e.g. the zebrafish *Danio rerio* or the fruitfly, *Drosophila*) or in yeast (a fungus), and many variations of the method have been applied. The resulting organisms can be important to understand normal cell or tissue functions throughout life or disease processes. In addition to studying protein function, transgenic organisms with inserted genetic material are extremely valuable to study the locations of proteins, or to express non-native 'tracer' proteins to detect particular cell types in complex, living organisms. Figure 5.2 shows an example of the expression of a green fluorescent protein in endothelial cells of the zebrafish, which allows the positions of blood vessels to be traced precisely. These genetically engineered organisms are valuable for testing and identifying new chemical compounds that may affect blood vessel patterning or function and may have future value to treat human diseases.

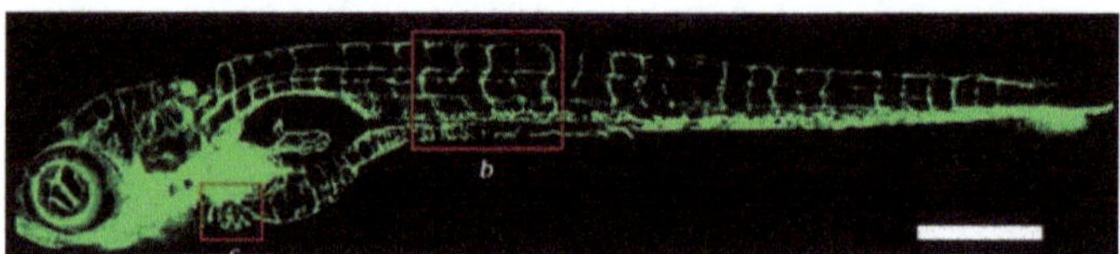

Fig. 5.2 A larval, transgenic zebrafish expressing green fluorescent protein in its endothelial cells to trace the patterning of blood vessels. Boxed area highlights vasculature in the liver. Scale = 500 microns. (Reproduced from Lawson and Weinstein 2002 [open access article])

5.3 Herbicide-Resistant Plants

Modern agriculture depends on use of many herbicides or pesticides. These poisonous chemicals kill weeds, fungi and other organisms that would otherwise reduce the crop yields (Preston 2019). It follows that the desired crop plants should be resistant to their poisoning action. Before the introduction of genetic engineering, herbicide-resistant plants were obtained by selective breeding over time between the variants of a particular crop species, or between different species. This is a slow and uncertain process. Companies involved in agribusiness have now developed genetic engineering methods in plants to generate transgenic plants and achieve efficient resistance against herbicides. An example is the glyphosate-resistant maize of the company Monsanto (recently purchased by Bayer). Glyphosate (Fig. 5.3) is a herbicide which kills essentially all natural plants through inhibition of a common enzyme. It now makes up around 25% of global herbicide sales.

This compound is used for crop protection in conjunction with herbicide-resistant crops. For example, Monsanto (now Bayer-Monsanto) sells glyphosate in a package deal with seeds of glyphosate-resistant maize. This policy and other decisions of the company have been much discussed in politics and public media. The positive feature has been a large increase in crop yield supporting food supply for humans. The negative features were that farmers became very dependent on the company and alternative ways to grow maize ceased. The new method is now widely used in North America and Canada and involves production of many thousand tons of glyphosate worldwide. In 2018, several institutions including Global 2000 and Greenpeace initiated a public vote in which more that 1 million European citizens applied to the European Union to ban use of glyphosate in EU member countries. The reasoning was based on a potential cancer-promoting action in humans (reviewed by Zhang et al. 2019) Glyphosate was classified in 2015 by the International Agency for Research on Cancer as "probably carcinogenic to humans". Although national health agencies consider on the current evidence that glyphosate is safe, in 2019 a court in California awarded two billion dollars to two individuals who claimed that they had developed non-Hodgkin lymphoma as a result of using, over years, a Monsanto herbicide based on glyphosate. A judge later reduced the award to $87 million (https://www.latimes.com/business/story/2019-07-26/monsanto-roundup-cancer-lawsuit-award). There is also evidence for other negative environmental effects, including possible effects on bees that are essential pollinators. For example, the gut bacteria of bees also contain

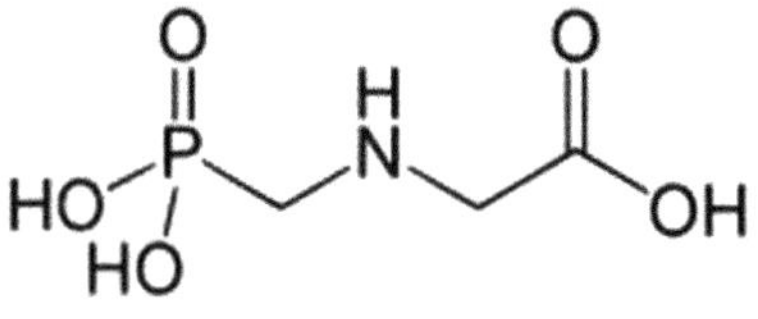

Fig. 5.3 The universal plant killer glyphosate. It acts by inhibition of the enzyme 5-enolpyruvylshikimate-3-phosphate synthase

the target enzyme, and if these bacteria are killed, the bees become more susceptible to infections (Motta et al. 2018). The decision made by the European Union was to extend the current permission to use glyphosate for 5 years (https://ec.europa. eu/food/plant/pesticides/glyphosate_en). In view of the uncertainties over effects of glyphosate on humans and other animals and the emergence of glyphosate-resistant weeds (Heap and Duke 2018; Nandula 2019), it is a concern that this new type of agribusiness continues to expand at a very high rate. The method has been expanded to other vegetables and fruits and glyphosate is also widely used for weed control in vineyards, which may affect microbes of the soil (Chou et al. 2018). Recently, large amounts of herbicides were found as contaminants in the drinking water in Switzerland (for example, https://www.swissinfo.ch/eng/chlorothalonil_banned-pes ticide-found-in-swiss-drinking-water--/45542188). A public initiative has been put in place to stop their use in agriculture. Given the issue of the development of resistant weeds, which will occur with extensive use of any chemical herbicide over time, other methods of weed control are under investigation. These include the engineering of more precise mechanical weeding machines, or the development of robots that could kill weeds with pulses of lasers or electricity (Stokstad 2019).

5.4 Cloning of Animals

It has long been a point of fascination for developmental biologists and others that complex, multicellular animals derive from a single, fertilised egg cell. Thus, the fertilised cell carries within itself the capacity or information to set up all the distinct differentiated cell types of the adult animal, a property referred to pluripotency. However, once differentiation to a specific cell type (for example, skin, nerve or muscle cells) has taken place, a very stable cellular state is reached that persists over years throughout the adult lifespan. By studying early embryos from animals such as frogs or starfish, developmental biologists showed that the cells of the very early embryo (termed the blastula stage of animal development) retain pluripotency and are essentially immortal. Later in animal development, the differentiation potential of different sets of embryonic cells becomes restricted to particular cell lineages that correspond to different tissue layers of the embryo. Stem cells of these lineages are referred to as "multipotent" (reviewed by Betschinger 2017). The different types of multipotent cells then give rise to specific cell types in the adult organism (Fig. 5.4).

Once scientific research of the twentieth century had established that DNA is the genetic material and resides within the nucleus of a eukaryotic cell (see Chap. 4), a major question emerged as to whether the restriction of the differentiation capacity of cells during embryonic development also involves a restriction of the genetic information available to the cells. For example, can the nucleus of a muscle cell only transcribe muscle-specific gene products, or can this nucleus ever be "re-programmed" to a multipotent or pluripotent state? This question is of deep scientific interest and has had very important implications for developing "designed" animals through cloning methods.

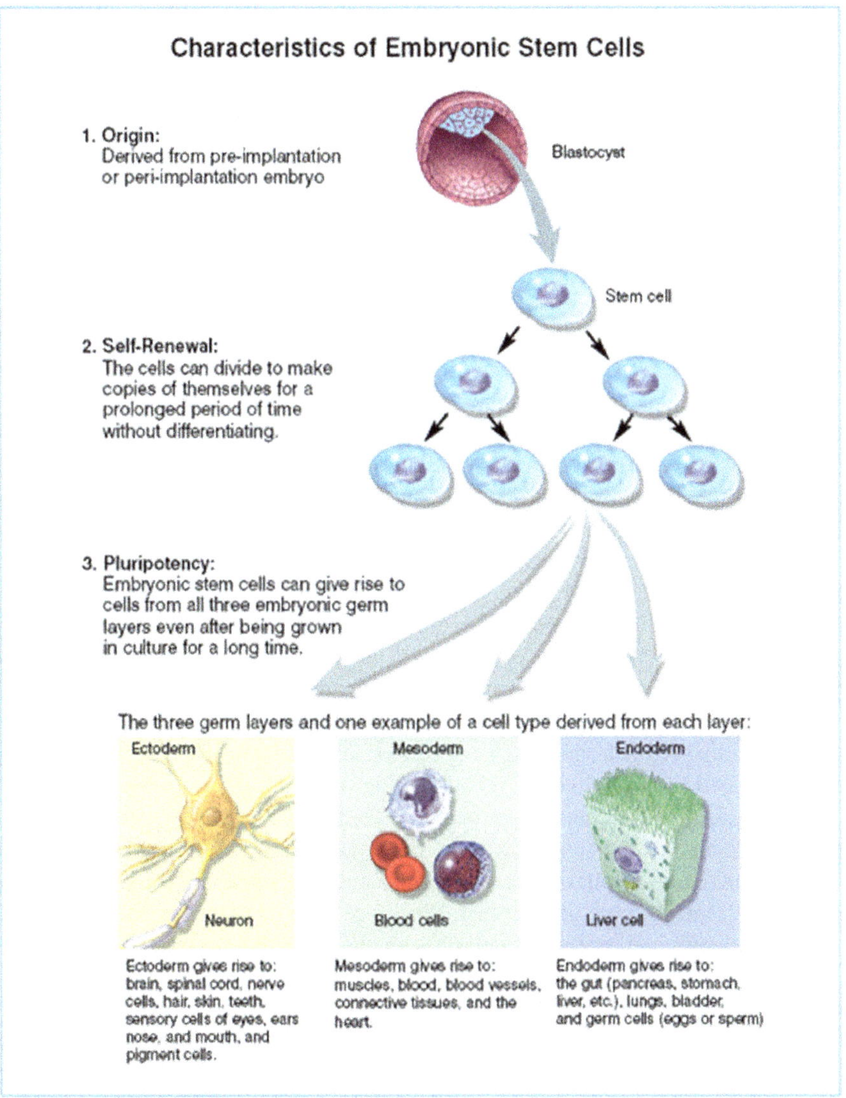

Fig. 5.4 Schematic of embryonic stem cells of animals and their relationship to multipotent cells and adult cell types. From NIH Stem Cell Information Home Page. In Stem Cell Information [World Wide Web site]. Bethesda, MD: National Institutes of Health, U.S. Department of Health and Human Services, 2016 [cited February 20, 2020]. Available at https://stemcells.nih.gov/info/Regenerative_Medicine/2006Chapter1.htm

To address this question, methods of nuclear transplantation were developed in the 1950s. The goal was to remove the nucleus of an egg cell, in a process termed "enucleation", and transfer a foreign nucleus into the cytoplasm of the egg cell. Pioneering studies by Briggs and King (1952), developed a method for using very thin glass needles to remove the nucleus of an amphibian egg cell and replace it with a nucleus from a blastocyst cell. The removal of the egg nucleus was found to prevent embryonic development, whereas egg cells that had been injected with a nucleus from a blastocyst cell continued with development. Indeed, 74% of these embryos completed embryonic development. This method set the scene for experiments involving nuclear transplantation from cells of different developmental stages. These culminated in experiments by Sir John Gurdon, when a post-graduate in Cambridge, UK, in which adult frogs were successfully derived (from 1.5% of the transplants) from enucleated eggs cells transplanted with nuclei from fully differentiated intestinal epithelial cells of late stage tadpoles (Gurdon 1962). Several years later, these frogs were shown to be fertile, providing another proof of the full reprogramming of the nuclei from the intestinal epithelial cells. In 2012, Sir John Gurdon was awarded the Nobel Prize in Physiology or Medicine for his discovery that mature, differentiated cells can be reprogrammed to become pluripotent. A commentary and perspective on the early papers can be found in a 2013 article (Gurdon 2013).

While experiments with amphibian embryos provided a productive and ethically acceptable route to exploring pluripotency, a major area of interest in this field relates to the cloning of higher vertebrates for experimental or practical purposes: examples are the laboratory mouse or farm animals. The first cloned mammal was a sheep, developed from a enucleated egg cell transplanted with a nucleus from a sheep embryonic cell at Cambridge University (Willadsen 1986, 1989). In 1997, the Roslin Institute in Scotland announced the origin of Dolly the sheep as the first mammal to be cloned from an adult somatic cell (Wilmut et al. 1997). Dolly was cloned by the introduction (by injection) of a nucleus from a mammary cell of a six-year old sheep into an enucleated oocyte of a Scottish Blackface sheep. During her life, Dolly gave birth to six lambs and received regular health checks. She was euthanised at the age of six because of lung tumours: in view that Dolly was a one-time experiment, it is not known if her clonal origin could have made her more susceptible to cancer.

In subsequent years, much more has been learnt about the reprogramming of adult cells, through the development of induced pluripotent stem cells (iPSC). This methodology was discovered by Shinya Yamanaka, while a member of the Institute for Frontier Medical Sciences, Kyoto University, Japan (Takahashi and Yamanaka 2006). Yamanaka subsequently shared the 2012 Nobel Prize for Physiology or Medicine with Sir John Gurdon. In the iPSC technology, cells from adult animals are reprogrammed by expression of four transcription factors (proteins that bind to DNA to control the expression of genes), namely, Oct4, Sox2, Klf4 and c-Myc. Each of these nuclear proteins controls the expression of specific sets of genes such that, when co-expressed, the host cell gains a pluripotent capacity similar to that of early embryonic cells. Major advantages of this methodology are that it can be applied to a huge range of species and cell types, and that it does not pose the complex ethical issues of using cells from early mammalian embryos. The applications of human iPSC for basic science research into developmental or disease processes, or for drug-screening and bioengineering, have grown very rapidly and offer great promise for the future of a new field of medicine, termed regenerative medicine (Shi et al. 2017).

5.5 Gene Editing Methods

Even since the debut of iPSC technology, breakthroughs and new frontiers in genetic engineering have continued. The ability to make genetic modifications to cells and organisms has been hugely expanded by the inception of a suite of technologies referred to as "gene editing". Previous efforts to introduce specific genetic mutations into target genes of vertebrates relied on lengthy methods of low efficiency and relatively high cost to generate mutations in the genome of an embryonic stem cell and then re-introduce these cells into blastocyst stage embryos. Instead, gene editing offers a straightforward way to target specific sites in the genome and to introduce changes, insertions, or deletions in the DNA sequence. A prominent method of gene-editing is based on the CRISPR/cas9 system, which stands for clustered regularly interspaced short palindromic repeats (CRISPR) and CRISPR-associated proteins (Cas). CRISPR/cas9 mediates a form of adaptive immunity in bacteria and archaea in response to invasive nucleic acids from viruses or plasmids (Sampson and Weiss 2014). By incorporating a DNA sequence from the invasive nucleic acid entity into the CRISPR sequence array of its own genome, a bacterium can then express crispr-repeats RNAs (crRNA) that hybridise to cognate sequences of the invader's DNA. This guides the Cas effector complex to target and cleave that DNA by endonuclease enzymatic activity, causing degradation of the invading viral or plasmid DNA (Barrangou et al. 2007).

Researchers have adapted this amazing biological process to work in the laboratory setting, such that any short region of any genome can be targeted by an engineered crRNA/Cas to achieve designed mutations. Since this technology can be applied to organisms that are not in use for laboratory experiments but for which the genome sequence is known, gene-editing has opened up a vast scope of organisms that were not previously accessible for molecular biology experiments. It should be noted that a concern with the Cas machinery is that it is possible for off-target DNA hybridisation to occur, meaning that unwanted mutations could also be introduced elsewhere in a genome. Thus "rescue" experiments—i.e., replacing of the wild-type sequence of the targeted gene—have an important role to demonstrate that any change in the properties of the cell or organism under study indeed results from the designed mutation. More recently, the specificity of the method has been improved and also new technologies such as "base editing" and "prime editing" have been developed, and a number of clinical trials are ongoing (Broeders et al. 2020). It seems plausible that gene editing, in some form, may be suitable for clinical applications in the future. However, a possible serious issue is that of immune reactions to the introduced Cas proteins (Broeders et al. 2020), which may limit the applications.

The CRISPR/Cas9 gene editing system is also being applied to plants and is seen by some as having potential to lead to new forms of agriculture. The new methods allow for well-directed changes of a plant genome. Under USA law, the modified plants are not classified as genetically modified species. This has led to new companies, such as Pairwise, a start-up company in Durham, North Carolina (https:// pairwise.com) that aims to use base-editing to modify berries and other products of agriculture to improve taste, colour and other properties and to respond to customer requests. The head of the company, Brian Crawford, is a former member of Monsanto who believes that world-wide agriculture will be completely changed by these new gene editing methods.

The cloning and genetic manipulation of animals is a controversial issue, with different legal frameworks in effect in different countries. In Switzerland it is forbidden, whereas the UK is more liberal, although all manipulations are carefully controlled. Although most countries have very tight regulation for all forms of genetic engineering of vertebrate animals, the genetic manipulation of primates remains ethically and legally possible. The first genetic engineering of a primate was carried out on rhesus monkeys in 2001. A passive "reporter gene" encoding the green fluorescent protein was introduced as a transgene into oocytes and a monkey was shown to express the green fluorescent protein in its cells, demonstrating the feasibility of the method (Chan et al. 2001). Since that time, several models of human disease, such as Huntington's disease, have been established in monkey species (Sasaki 2015).

Although there is a value to model a disease process in a primate species, which is genetically much closer to humans than mice or rats, experiments with primates incur many major ethical issues given the high cognitive abilities and emotional capacity of primates. In some countries, the primates most closely related to humans (the great apes: gorillas, orangutan, bonobos and chimpanzees) are given specific legal protections and/or the use of these species is banned for experimentation. There is also ongoing activity and debate in the USA and other countries as to whether the great apes should be recognised legally as "persons", i.e., as autonomous living entities with capacities for rights of their own. In Argentina in 2014, a landmark case awarded specific legal rights to an orang-utan to be transferred to a new habitat (https://edition.cnn.com/2014/12/23/world/americas/feat-orangutan-rights-ruling/).

5.6   Designed Humans

In 2018, researcher Dr. Jiankui He in China publicised "designed" human babies at the second international summit of human genome editing (Fig. 5.5). He claimed to have worked with clinicians to produce genetically edited babies at the Southern University of Science and Technology in Shenzhen, East China. The genomes of the embryos were changed by the CRISPR/cas9 gene editing technology, and the embryos then implanted into two women. As announced in November 2018, one of the women gave birth to natural twins named Nana and Lulu. The family medical history was that the children's mother is healthy whereas her husband suffers from

Fig. 5.5 Jiankui He at the second international summit of human genome editing in October 2018

AIDS, a serious disease that can result from infection by the HIV virus. The goal of the gene editing was to eliminate the cell-surface protein used by the HIV virus to enter its target cells. The researchers' claim for the experiment was that the children will now be resistant to HIV infection, and thereby AIDS, throughout their lives.

Jiankui He and his collaborators published their results only in public media (https://www.youtube.com/channel/UCn_Elifynj3LrubPKHXecwQ/videos). and not by the academic process of publication in a peer-reviewed scientific journal. Severe criticisms of his work in the scientific community and in public media have ranged from doubts of whether the account of the experiments is true, to fierce condemnation of the complete lack of scientific responsibility shown by carrying out such an experiment. The main area of criticism was that this type of experiment should not have been performed. As we discussed in Sect. 5.5, it is generally accepted that the genetic change introduced by the scissor-like action of CRISPR/cas9 is not entirely error-free, i.e., the embryo's genome might have been cut and changed elsewhere than the planned site. In this way, additional genetic changes would be produced and potentially carried through future generations.

Compared with these very major ethical issues and risks, the medical advantage of obtaining AIDS resistance is small and could be achieved by other methods that do not involve unknown risks and potentially hereditable genetic changes. The University of Shenzen itself declared that it did not know about the experiment of Dr. He. The newspaper "China Daily" reported that he did not ask for an ethical evaluation from the University or funding authorities. In December 2019, He and two associates were found guilty by a court in Shenzhen of "illegal medical practices". These included the forging of ethical review documents, misleading medical doctors, and deliberately violating Chinese biomedical research regulations. All three researchers received large fines and prison sentences, with He receiving a three-year sentence (Normille 2019). Since this case, the Chinese government has increased the regulations around human gene editing.

It remains possible that the general ethical consensus around genetic engineering might be changed in the future for genetic manipulations for which a medical advantage is convincing. Clearly, there are many debilitating natural genetic diseases of humans, for example, cystic fibrosis, which could in theory be rendered non-inheritable by genetic editing in the germ line of humans known to carry a cystic fibrosis mutation (see Chap. 4, Sect. 4.3 for examples of mutations that cause cystic fibrosis). However, multiple genetic engineering approaches to cystic fibrosis are under development or in clinical trial and the majority of these do not require alterations to the germ line. Instead, they focus on restoring normal cystic fibrosis trans-membrane conductance regulator channel function specifically in the airway epithelium (Boyd et al. 2019). If the methods in clinical trial prove successful, there will be no need to consider alterations to the germ line to control this disease. It remains possible that genetic engineering in the germ line might be permitted in the future, with regard to positive effects on ameliorating genetic diseases that are life-threatening or have major effects on quality of life and for which there are no other possible treatments. Against this, are the strong ethical issues and high risks that include starting a possible artificial evolution in the human population and also the danger of disastrous misuse for non-health related purposes.

5.7 Artificial Life Has Not Yet Been Achieved, but Nomenclature Causes Confusion

In 2019, a Basel daily newspaper published a detailed review by Christoph Bopp (2019) with the sensational title "The totally artificial bacterium". This review, which was supported by the Gebert Rüf Foundation, was based on a publication by Fredens et al. (2019) that described changing the natural genome of *E. coli* to a genome completely synthesized by methods of organic chemistry. The DNA synthesis method offers many advantages for genetic manipulations of organisms. Pioneering work in this field was performed by Gibson and colleagues (Gibson et al. 2008). The authors of this paper exchanged the wild-type genome of the bacterium *Mycoplasma genitalium* (which has a genome of less than 1 Megabase (Mb) of DNA, much smaller that the genome of *E.coli*) with a chemically-synthesised genome. The method was slow and very costly at that time but has now been improved in speed and has lower costs. The synthesis of larger complete genomes is now possible, as shown by the 2019 publication on *E. coli* that involved replacement of a 4 Mb genome (Fredens et al. 2019).

Prior to chemical synthesis of DNA, genetic manipulations by researchers were carried out principally by mutagenesis of a wild type gene by methods of molecular biology (see Sect. 5.1), or by random mutations induced by a chemical mutagen, or by selective breeding by genetic crosses. Thus, while the method for synthesis of a gene or genome is new, the same piece of DNA could have been obtained by molecular biology approaches. The correct title of the review by Bopp (2019) should be "A bacterium with a totally artificial genome". This difference displays the difficulty of distilling the essence of genetic engineering experiments to wider audiences.

The genome itself is not alive. To achieve a living cell, cell membranes, energy sources and cell structures are also needed. In the experiments of Fredens et al. (2019), all the other features of a natural, live *E. coli* cell were maintained. Only the parent genome was exchanged for the synthetic genome.

In fact, the work of Fredens et al. (2019) (as also reviewed by Blount and Ellis (2019), did not have the aim to create an artificial bacterium. The goal was to examine the effect of using different codon building blocks in its genome. In *E. coli,* the triplet of bases (the codon) TCG codes for the amino acid serine and the codon AGC is an alternative codon for the same amino acid. Within the synthetic genome, all the triplets TCG were replaced by ACG. Several additional changes were introduced in the synthetic *E. coli* genome so that it could be identified within the bacterial cells. This "bar-coded" genome was used to replace the wild-type genome in natural *E. coli* cells, giving rise to a modified bacterium named E.coliSyn61. The E.coliSyn61 bacteria grow somewhat more slowly (a 1.6-fold increase in population doubling time) than the natural *E. coli*, but in most aspects that were examined by the researchers E.coliSyn61 appears to have very similar properties to the parent strain (Fredens et al. 2019). E.coliSyn61 thus joins a wide range of genetically altered *E. coli* strains, but it is not a totally artificial bacterium.

Another new approach has been to use so-called evolutionary algorithms (algorithms that mimic biological mechanisms of mutation, recombination and natural selection to generate optimised designs according to pre-specified criteria) to shape novel assemblies of cells that are predicted to behave in certain ways. The concept is to take advantage of the bio-compatibility of natural living cells to generate small "living machines" that could potentially have many applications; for example, to deliver drugs within the body, move over distances, or clean up environmental contaminants. In a proof-of-principle study, the authors took advantage of the great self-renewal capacity of embryonic cells to isolate cells from early *Xenopus* embryos, culture them as aggregates, and then conduct micro-surgery to achieve the desired shape for the "xenobots" (Kriegman et al. 2020). Although these are indeed novel and artificial assemblies of cells, distinct from normal embryos or organs, the property of life in the cells comes from entirely natural *Xenopus* embryo cells and thus the xenobots, as they are at present, do not constitute artificial life.

Research into the possibility of generating artificial life is highly ambitious and connected to hopes of building a "new nature" to combat food shortages, environmental change, or other limitations, possibly designed under the guidance of artificial intelligence. The field includes much speculative thinking. It has even been proposed that artificial organisms composed of inorganic compounds could be generated (see Chap. 9). We feel that this is very unlikely in view of the much lower versatility of inorganic chemistry (see Chap. 3, Sects. 3.4 and 3.3). As discussed in Chap. 3, the generation of precursors of life on the basis of organic chemistry appears to be possible, but how to achieve this in the laboratory has not been solved experimentally. The experimental generation of life at the level of a complete bacterium or a more complex organism seems at present to be extremely unlikely.

References

Barrangou R, Fremaux C, Deveau H, Richards M, Boyaval P, Moineau S, Romero DA, Horvath P (2007) CRISPR provides acquired resistance against viruses in prokaryotes. Science 315(5819):1709–1712

Berg P (2008) Asilomar 1975: DNA modification secured. Nature 455:290–291

Betschinger J (2017) Charting developmental dissolution of pluripotency. J Mol Biol 429(10):1441–1458

Blount BA, Ellis T (2019) Construction of an Escherichia coli genome with fewer codons sets records. Nature 569(7757):492–494. https://doi.org/10.1038/d41586-019-01584-x

Blount ZD (2015) The natural history of model organisms: the unexhausted potential of *E. coli*. eLife 4:e05826. https://doi.org/10.7554/eLife.05826

Bopp C (2019) Das total künstliche Bakterium. CH Media (Wissenschaft bewegen, Gebert Rüf Stiftung) in BZ (Basler Zeitung) May 11, 2019

Boyd C, Guoc S, Huangc L, Kerem B, Orene YS, Walkerf AJ, Hartf SL (2019) New approaches to genetic therapies for cystic fibrosis. J Cyst Fibros. https://doi.org/10.1016/j.jcf.2019.12.012 (aheadoffinalissuepublication)

Briggs R, King TJ (1952) Transplantation of living nuclei from blastula cells into enucleated frogs' eggs. Proc Natl Acad Sci U S A 38(5):455–463. https://doi.org/10.1073/pnas.38.5.455

Broeders M, Herrero-Hernandez P, Ernst MPT, van der Ploeg AT, Pijnappel WWMP (2020) Sharpening the molecular scissors: advances in gene-editing technology. iScience 23(1):100789. https://doi.org/10.1016/j.isci.2019.100789. Epub 19 Dec 2019

Buchman A, Gamez S, Li M, Antoshechkin I, Li H-H, Wang H-W et al (2020) Broad dengue neutralization in mosquitoes expressing an engineered antibody. PLoS Pathog 16(1). https://doi.org/10.1371/journal.ppat.1008103

Carvalho DO, McKemey AR, Garziera L, Lacroix R, Donnelly CA, Alphey L, et al (2015) Suppression of a field population of Aedes aegypti in Brazil by sustained release of transgenic male mosquitoes. PLoS Negl Trop Dis 9:e0003864

Chan AWS, Chong KY, Martinovich C, Simerly C, Schatten G (2001) Transgenic monkeys produced by retroviral gene transfer into mature oocytes. Science 291:309–312

Chou MY, Vanden Heuvel J, Bell TH, Panke-Buisse K, Kao-Kniffin J (2018) Vineyard under-vine floor management alters soil microbial composition, while the fruit microbiome shows no corresponding shifts. Sci Rep 8(1):11039. https://doi.org/10.1038/s41598-018-29346-1

Engel J (1992) Laminin and other strange proteins. Biochemistry 10:10643–10651

Engel J (2017) A critical survey of biomineralization. Springer Briefs in applied Science and Technology, ISSN 2191–530X, pp 1–64

Fredens J, Wang K, de la Torre D, Funke LFH, Robertson WE, Christova Y, Chia T, Schmied WH, Dunkelmann DL, Beránek V, Uttamapinant C, Llamazares AG, Elliott TS, Chin JW (2019) Total synthesis of Escherichia coli with a recoded genome. Nature 569:514–518

Heap I, Duke SO (2018) Overview of glyphosate-resistant weeds worldwide. Pest Manag Sci 74(5):1040–1049. https://doi.org/10.1002/ps.4760

Gibson DG, Benders GA, Andrews-Pfannkoch C, Denisova EA, Baden-Tillson H, Zaveri J, Stockwell TB, Brownley A, Thomas DW, Algire MA, Merryman C, Young L, Noskov VN, Glass JI, Venter JC, Hutchison CA 3rd, Smith HO (2008) Complete chemical synthesis, assembly, and cloning of a Mycoplasma genitalium genome. Science 319:1215–1220

Gurdon JB (1962) The developmental capacity of nuclei taken from intestinal epithelium cells of feeding tadpoles. J Embryol Exp Morphol 10:622–640

Gurdon JB (2013) The Cloning of a Frog. Development 140(12):2446–2448. https://doi.org/10.1242/dev.097899

Gutekunst J, Andriantsoa R, Falckenhayn C, Hanna K, Stein W, Rasamy J, Lyko F (2018) Clonal genome evolution and rapid invasive spread of the marbled crayfish. Nat Ecol Evol 2(3):567–573. https://doi.org/10.1038/s41559-018-0467-9

Kriegman S, Blackiston D, Levin M, Bongard J (2020) A scalable pipeline for designing reconfigurable organisms. PNAS 117:1853–1859. https://doi.org/10.1073/pnas.1910837117

Lawson ND, Weinstein BM (2002) In vivo imaging of embryonic vascular development using transgenic zebrafish. Dev Biol 248:307–318

Motta EVS, Raymann K, Moran NA (2018) Glyphosate perturbs the gut microbiota of honey bees. PNAS 115:10305–10310. https://doi.org/10.1073/pnas.1803880115

Nandula VK (2019) Herbicide resistance traits in maize and soybean: current status and future outlook. Plants (Basel) 8(9):337. https://doi.org/10.3390/plants8090337

Normille D (2019). News report in Science 367. https://doi.org/10.1126/science.aba7347

Preston C (2018) The synthetic age. MIT Press

Preston C (2019) Sind wir noch zu retten? Wie wir mit neuen Technologien die Natur verändern können. Springer

Sampson TR, Weiss DS (2014) Exploiting CRISPR/Cas systems for biotechnology. Bioessays 36(1):34–8. https://doi.org/10.1002/bies.201300135

Sasaki E (2015) Prospects for genetically modified non-human primate models, including the common marmoset. Neurosci Res 93:110–115. https://doi.org/10.1016/j.neures.2015.01.011

Shi Y, Inoue H, Wu JC, Yamanaka S (2017) Induced pluripotent stem cell technology: a decade of progress. Nat Rev Drug Discov 16(2):115–130. https://doi.org/10.1038/nrd.2016.245

Stokstad E (2019) Jury verdicts cloud future of popular herbicide. Science 364:717–718

Takahashi K, Yamanaka S (2006) Induction of pluripotent stem cells from mouse embryonic and adult fibroblast cultures by defined factors. Cell 126(4):663–676

Willadsen SM (1986) Nuclear transplantation in sheep embryos. Nature 320:63–65

Willadsen SM (1989) Cloning of sheep and cow embryos. Genome 31:956–962

Wilmut I, Schnieke AE, McWhir J, Kind AJ, Campbell KH (1997) Viable offspring derived from fetal and adult mammalian cells. Nature 385:810–813

Zhang L, Rana I, Shaffer RM, Taioli E, Sheppard L (2019) Exposure to glyphosate-based herbicides and risk for non-Hodgkin lymphoma: a meta-analysis and supporting evidence. Mutat Res 781:186–206. https://doi.org/10.1016/j.mrrev.2019.02.001

Chapter 6
Natural Risks to Life

Throughout the history of life on Earth, living organisms have been exposed to many threats of different calibres. In modern texts, disasters are grouped into either natural or human-made risks. Of the natural risks, in this chapter we will consider risks posed by epidemics and pandemics, and then move onto localised severe climatic or geological events. Finally, we will focus on extra-terrestrial risks from asteroids or other cosmic bodies that may hit the Earth. There is strong and growing evidence that the presently observed, rapid increase in global average temperature, called global warming, is a driver of severe climate change and is due to human activities. This topic will be covered in Chap. 8.

6.1 The Microbial World, Epidemics and Pandemics

Each species on Earth faces a large number of natural risks, not least from ecological interactions and predation between living organisms. As described in Chap. 3, only photosynthetic organisms can convert energy directly from sunlight into chemical energy. Many organisms instead obtain their nutrients by ingesting or parasitizing other living organisms. The web of life on earth has therefore evolved complex networks of predator-prey interactions. In a stable environment, these interactions can exist in a steady-state or equilibrium in which the predators do not kill all the prey. However, the ratio of predators to prey can alter if the environment changes: if predators become very numerous this will lead to a decrease in numbers of a prey species. This will result in a local food shortage for the predators, further leading to a decline in the number of predator individuals. Reduced predation will enable the numbers of prey to increase again. These relationships were mathematically analysed by Glansdorf and Prigogine (1971). As an example, when humpback whales chase shoals of herrings and consume large numbers of fish, this does not pose an ecological problem as long as a steady state exists and herrings as well as whales remain in existence.

J. C. Adams and J. Engel, *Life and Its Future*,
https://doi.org/10.1007/978-3-030-59075-8_6

Other interactions between organisms involve the microbial world, which includes viruses, bacteria, fungi, protozoa and helminths. These comprise millions if not billions of species and are the most numerous forms of life on earth. For bacteria alone, it is estimated that there are a total of 5×10^{30} bacteria on Earth (Nature Reviews Microbiology 2011), potentially including an estimated 0.8×10^6 to 1.6×10^6 taxonomically diverse forms (Louca et al. 2019). These collectively occupy a huge range of ecological niches. Most forms of life exist in association with microbes and the microorganisms may in fact have beneficial effects on their host. For example, humans carry ~ 1 kg of bacteria within the digestive tract (referred to as the microbiome or microbiota), which assist in the breakdown of food, immunity, protection against pathogenic microbes and other functions. These functions of the microbiota may have helped humans adapt to different diets as the human population spread from Africa around the world (see Chap. 4) (Suzuki and Ley 2020).

Virus particles are a special category of microbe, which are not in themselves alive. Viruses have single- or double-stranded genomes based on either DNA or RNA and the genome is encased in a lipoprotein particle (Duffy 2018). Viruses depend absolutely on their ability to infect a host cell for their reproduction. Once within the host cell, the virus reprograms the cell's protein synthesis apparatus to produce virus-encoded proteins and replicate the viral genome, so that new virus particles can be released from the cell. It is not clear how viruses evolved, but the current scenario is that viruses originated at the same time as the first proto-cells, followed by new groups of viruses that have emerged repeatedly over time (Krupovic et al. 2019), with differing trophisms to infect different types of host cells. Bacterial cells themselves are colonised by viruses known as bacteriophages. It is estimated that around 10^{31} viral particles exist on Earth (discussed by Mushegian 2020).

In many cases, microorganisms associate with their host species with beneficial effects, or little effect, on the health of the host. However, there are also many microorganisms that cause devastating diseases. Some of these are transmitted to the host by an intermediate species (the vector or intermediate host). For example, malaria is a tropical disease of humans caused by the *Plasmodium* protozoan which is transmitted to humans by blood-sucking mosquitos. Malaria causes around 0.4 million deaths worldwide each year as calculated by the World Health Organisation (WHO). Pandemics (epidemics that spread around the world) caused by viruses and bacteria have been frequent events in human history and were reported from the earliest times, long before the causative agents were known. The Black Death (bubonic plague), caused by the bacterium *Yersinia pestis*, appears to have caused the most deaths. It originated in Central Asia and reached Europe through rats and fleas that infested Genoese merchant ships. In Europe, the plague killed between 30% and 60% of the population between 1347 and 1351, an estimated 200 million victims. Smallpox is another deadly disease that frequently ravaged Europe and is thought to have originated in prehistoric times. The Antonine plague of 165–180, which killed about 5–7 million people and contributed to the decline of the Roman Empire, is thought to have been due to smallpox. Smallpox was unknown in the Americas until accidently introduced by the Spanish and Portuguese conquistadors, whereupon it contributed to the termination of the Aztec and Inca empires. Smallpox epidemics continued in Europe over centuries, including great Plague of London of 1665.

The first scientific vaccine trials against smallpox were carried out by Edward Jenner, an English doctor and surgeon with a strong interest in natural history. Through his country medical practice, Jenner was aware that milkmaids had protection from smallpox after having caught a milder disease, cowpox, from the cows they milked. In 1796, he took fluid from the skin pustules of Sarah Nelmes, a dairymaid who had contracted cowpox, and inoculated a healthy eight-year old boy, James Phipps, the son of his gardener (Fig. 6.1). After initial discomfort, James continued healthy. One month later, Jenner inoculated James with fluid from a fresh smallpox lesion, yet the boy did not develop smallpox. Jenner went on to inoculate and test twenty-three other individuals and wrote up his results in a paper published privately in 1798. He promoted the concept of vaccination internationally and made his cowpox vaccine available to other doctors and researchers. He vaccinated local people for free in a "Temple of Vaccinia" built in his garden. Jenner was widely honoured during his lifetime but tragically lost several family members to another infectious disease, tuberculosis, and died of a stroke in 1823 (Riedel 2005).

Throughout these times, the causative agents of infectious diseases were unknown. In 1861, Louis Pasteur revolutionised medical thinking by putting forward his germ theory of human disease, which proposed that pathogenic microorganisms (germs) were transmitted from an infected person to other people and were responsible for human epidemic diseases (Berche 2012). During the nineteenth and early twentieth centuries, social improvements to sanitation, lifestyle and general health, along with scientific advances leading to vaccination programmes, led to dramatic decreases in mortality from infectious diseases such as smallpox, typhoid and cholera in developed countries. Great improvements in nursing care were pioneered by Florence Nightingale (1820–1910), building from her experience in combating infectious diseases in a British military hospital during the Crimean War. Nightingale emphasised good hygiene, regular hand-washing and evidence-based medical practices and in 1859

Fig. 6.1 A painting representing Edward Jenner's first inoculation of James Phipps on 4th May 1796. Image from Wikipedia, English version

she was the first woman elected to the Royal Statistical Society. Her legacy remains relevant to infection control today (Ellis 2020).

Within the twentieth century, further steps against infectious diseases were taken with the discovery of antibiotics. The means to produce these agents in bulk greatly reduced the lethality of diseases caused by bacteria. By the 1960s, the success of antibiotics and the continuing decline in deaths due to infections led to an optimistic point of view that human infectious diseases had been conquered. However, since the 1980s, many new infectious diseases have appeared including, amongst others, Legionnaire's disease, HIV/AIDS, Lyme disease and sudden acute respiratory syndrome (SARS). Bacterial and viral diseases have increased decade on decade as monitored by the World Health Association. At present, around 20% of deaths worldwide are caused by infectious diseases. In parallel, widespread overuse of antibiotics has unfortunately taken place and there has been no development of new antibiotics since the 1980s. These conditions have fostered the evolution of antibiotic-resistant bacteria, especially in hospitals, to the extent that many infections are no longer treatable with the current antibiotics. It is estimated that 700,000 people now die globally each year because of antibiotic resistant bacteria (https://www.antibi oticresearch.org.uk/about-antibiotic-resistance/). The development of new types of antibiotics has become an extremely urgent goal of pre-clinical research (Smith et al. 2015).

The American microbiologist and Nobel Prize winner Joshua Lederberg defined that the continuing success of medical treatments must address the issue of the natural evolution of microbes. He warned of an "evolutionary arms race", pointing out "we cannot compete with microorganisms whose populations are measured in exponents of 10^{12}, 10^{14}, 10^{16} over periods of days". "In the race against microbes our best weapon is our wits, not natural selection on our genes" (Lederberg 1997). In general, the human immune system responds to antigenic features of the surface of bacteria or viruses. Mutations that alter surface proteins, so that the immune system is presented with novel antigens, may give the pathogen a biological advantage of transmission, and/or result in a weaker immune response, potentially increasing the infection rate. This means that a new mutant variant of a virus or other pathogen can lead to an epidemic and, in the worst cases, a worldwide pandemic.

This was seen with the so-called Spanish flu (1918–1920). Influenza is cased by a virus which has an RNA genome. These viruses undergo a higher rate of mutations than DNA viruses and so various strains of flu virus are in circulation. The Spanish flu pandemic started at the end of the First World War, at a time when most people suffered from hunger and many troops were returning to different countries. It was caused by a type A influenza virus of subtype H1N1, which was a new mutant strain particularly lethal to younger people. It is estimated that the 1918 flu variant killed between 50 and 100 million people, many more than the number of victims of the war itself. Nowadays, in most years, influenza viruses and the viruses that cause the common cold (typically rhinoviruses) infect many people, but the majority of people have some prior immunity and recover within a few days. Because viral infections cannot be treated with antibiotics, vaccination is the general approach to therapy. For example, in many countries, flu vaccines are routinely offered each winter. However, the effectiveness of a flu vaccine cannot be guaranteed because the flu virus mutates rapidly (a feature of viruses with RNA genomes) and it is unpredictable which flu subtypes may predominate each winter.

At the time of writing this book, humans are suffering from a global pandemic due to a novel coronavirus named SARS-CoV-2 that has killed over 3.75 million people globally to date. This virus causes a respiratory disease termed covid-19 (corona virus disease-19) (Zhou et al. 2020). Other types of coronavirus can cause the common cold. Since humans have never experienced the SARS-CoV-2 virus before, at the start of the epidemic no person had immunity to the new virus. The virus is often lethal in older people and in those who have compromised immune systems or underlying health conditions. In spring 2020, to limit infections by SARS-CoV-2 while testing kits and vaccines could be developed, most countries closed all public events, schools, shops and most workplaces. People were advised to stay at home and work from home where possible. Many counties closed their borders and restricted public transportation. We can learn from earlier epidemics that these isolation measures will limit person-to-person transmission and the number of new cases should drop after a few months. However, opening up normal activities again whilst most of the population has no immunity will lead to further waves of cases. The second wave seen in many countries towards the end of 2020 has involved many more cases of covid-19 than the first wave, driven by the emergence of new, more transmissible variants of the virus.

The coronaviruses are RNA viruses and tracing of the outbreak was made possible by quickly developed tests for the RNA of SARS-CoV-2, which were set up by various pharmaceutical companies and other organisations. Most tests use a molecular biology method known as qRT-PCR on nasal or throat swabs to test for the presence of the viral RNA. From the sample, viral RNA regions are transcribed into complementary DNA. The DNA is amplified by a polymerase chain reaction for detection by a fluorescent signal. Some of these tests are hampered by a significant false negative rate, which greatly complicates the identification of people who may be infectious (Basu et al. 2020). However, these tests are very important in view that not all covid-19-infected people have symptoms of disease.

The development, manufacturing and distribution of effective reagents to test if a person who has been exposed to the virus has developed immunity takes longer to set up. A very urgent goal of 2020 was to develop vaccines to potentially protect people against the effects of further covid-19 infection. The new coronavirus and related coronaviruses are viral particles with some unique features (Fig. 6.2). The surface spike protein of SARS-CoV-2 attaches to the receptor angiotensin-converting enzyme II (ACEII) on target cells in the human respiratory tract and assists the insertion of the RNA genome into the target cells. Multiple vaccines for SARS-CoV-2 have now been developed with remarkable speed and success in clinical trials. Approvals for mass vaccination began in December 2020. Although Edward Jenner had the freedom to inject his trial subjects with smallpox virus to determine if they had developed resistance, in modern times such "challenge trials" are viewed as unethical, especially when there is no treatment available and the disease is not well understood. Nevertheless, the WHO has issued new ethical guidelines for challenge trials (Jamrozi et al. 2021). The first covid-19 challenge trial was launched in the UK in February 2021,

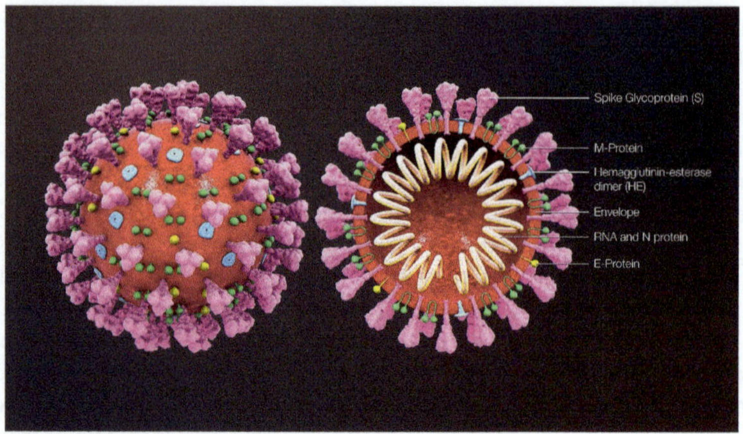

Fig. 6.2 Schematic structure of a coronavirus. The RNA genome is surrounded by an envelope consisting of a lipid membrane and inserted membrane proteins. The functional groups point outwards and the proteins are therefore called surface proteins. From Wikipedia, English edition 2020

to learn more about infection and re-infection from controlled covid-19 infection of healthy, young volunteers (https://www.gov.uk/government/news/worlds-first-cor onavirus-human-challenge-study-receives-ethics-approval-in-the-uk).

During the time that SARS-CoV-2 vaccines have been under development, a wide range of basic research on the virus has also started. In addition to understanding how it binds to human cells, we need to learn about the factors that make people susceptible to the most severe forms of covid-19. Here, genome sequencing has an important role to play. A region on human chromosome 3 that is carried by around 50% of Europeans and 16% of South Asians was found to be associated with risk of respiratory failure upon infection with SARS-CoV-2. This region centres on a 50 kilobase region inherited from the Neanderthals (Chap. 3) (Zeberg and Pääbo 2020). Another source of susceptibility appears to be genetic variants that affect a person's ability to produce type 1 interferon as a first defence against the viral infection, before an antibody response has been activated (Hadjadj et al. 2020). A study of UK-based patients has confirmed the association of the chromosome 3 region and identified several new genetic associations, including a variant of the gene that encodes the cell-surface receptor for interferon, IFNAR2 (Pairo-Castineira et al. 2020). It is likely that other genetic associations will be identified as more studies are carried out around the world. The results to date provide new insights on the mechanisms behind the disease and possible approaches to therapies.

Many questions remain on how people may safely emerge from isolation to resume a more normal way of life without triggering more waves of infections and the emergence of new viral variants that may escape current vaccines. The pandemic caused by SARS-CoV-2 is instructive for the enormous changes in public life and the economies of countries that have been caused by this global medical disaster.

As the virus spread from country to country, rapid mutational change of its RNA-based genome has allowed for real time tracking of its migration and evolution. An open-source platform, Nextstrain (nextstrain.org), developed by Hadfield et al. (2018), uses publically available genomic data to track pathogen infections and has identified several different strains of SARS-CoV-2 in different parts of the world. For example, new variants that transmit more effectively were identified in the UK and in South Africa. This is a concern for the effectiveness of current vaccines over time.

We may dislike the microbes that cause deadly diseases, but we have to recognise that they are part of the global biosphere. As discussed in Chap. 3, viruses originated as part of biological evolution. It follows from the observation of many human pandemics that the host population is never completely extinguished. Successful defence is possible because the human immune system adapts to all kinds of infections and has "memory", i.e., a stronger immune response is caused when a person becomes re-infected by the same or very similar infectious agent. Because of genetic variation in the human population there will also be some individuals who are less susceptible to the effects of the infectious agents, with so-called natural resistance. Although microbes are often very host-specific, by mutation they may acquire the ability to be transmitted from one species to another. In this situation, the microbe is unknown to the immune system of the new host and the primary immune response may not be sufficient to protect the new host from serious effects of the infection. A jump of species is what has happened with the SARS-CoV-2 virus, which is thought to have crossed to humans from bats or from other wild-animal species that are sold in food-markets in China. The SARS-CoV-2 genome sequence is 96% identical to that of a coronavirus that infects bats (Zhou et al. 2020). In humans, the virus is then transmitted from person to person by respiratory droplets. There are also concerns that the virus may also be spread by airbourne pollution (Domingo et al. 2020).

The inter-dependence of an infectious agent with its host and how they co-evolve under natural selection is illustrated by the long-term efforts to control the population of rabbits in Australia. The rabbit is a native species of Europe and eastern Asia that was taken around the world by human exploration, colonisation and migration and became a particular pest in countries that lack its natural predators. In Australia, by the 1930s, the rabbit population had reached an estimated 10 billion. The catastrophic effects of these rabbits eating so much of the native vegetation had major economic impacts on sheep farmers and led to the extinction of native species of plants and animals. In the 1950s, the Australian government approved a form of biological control on the rabbits: environmental release of the myxoma poxvirus. Myxoma virus is transmitted by biting insects and infects only rabbits and European brown hares to cause a fatal disease, myxomatosis. This was the first time that humans had deliberately introduced a virus into the wild. It is estimated that 400 million rabbits were killed in one year and within ten years the Australian rabbit population had decreased by 95%. However, over time, the virus and the rabbits co-evolved under natural selection for less virulent virus variants and rabbits that had genetic resistance to the virus. Within seven years, rabbit numbers stabilised again as mortality from myxomatosis decreased to 30% and myxoma became an endemic, non-fatal virus (Spiesschaert et al. 2011). Over time, the rabbit population began to increase again. A second biological control programme was started in 1996 based on a new disease of rabbits caused by a calicivirus, Rabbit Haemorrhagic Disease virus (RHDV). Release of RHDV was successful for over 10 years in decreasing the numbers of rabbits to around 200 million and allowing recovery of vegetation and native species. However, again due to co-evolution, the lethality of RHDV started to diminish and rabbit numbers began to increase again. Research projects in Australia trialled RHDV strains from different countries for more lethal variants and a RHDV-K5 strain from Korea was released in 2017 (Agriculture Victoria 2017). A research program to accelerate the evolution of the virus under laboratory conditions was also set up, with the goal of generating distinct myxoma strains that can be released over time to continually suppress the rabbit population (Hall et al. 2017).

We also emphasise that epidemics and pandemics are not specific to animals and also occur in plant species. These can have dramatic major consequences for ecosystems with effects on human health, food supply and livelihoods. The devastating Potato famine in Ireland (1846–1850) was caused by infection of potato crops by the fungus *Phytophthora infestans* and led to famine, the death of one million people and widespread emigration. Dutch elm disease is an example of a tree pandemic caused by a fungus which has killed billions of elm trees globally, leading to major restructuring of landscapes, ecosystems and biodiversity (Potter et al. 2011). Another devastating tree pandemic began in the early twentieth century, when the fungus *Cryphonectria parasitica*, which causes chestnut blight, was accidently brought to the USA from Japan. Over time an estimated 300 million native trees in the USA have been killed. The native American chestnut tree has been driven essentially to extinction, transforming habitats, ecosystems, human lifestyles and agricultural practices in the eastern USA (Smith 2000). Efforts continue to develop genetically resistant strains from the few remaining chestnut trees by selective breeding, or hybridisation with non-native chestnut species. A new approach is to introduce blight tolerance into the native chestnut species by engineering into its genome an oxalate oxidase-encoding gene from wheat. This enzyme breaks down oxalate produced by the fungus and so reduces its pathogenicity (Powell et al. 2019). The concept is that blight tolerance is less likely than blight resistance to lead to the emergence of new virulent strains of the fungus.

Just as in animals, a new epidemic or pandemic can emerge in plants because a microorganism transmits to a new host species. Ash die-back is a new disease that is devastating and killing ash trees in Europe, having emerged in Poland in 1992 and reached the UK by 2012. It is caused by a fungus, *Hymenoscyphus fraxineus*, that has jumped species from the Asian ash. The fungus has little effect on Asian ash, but only 5% of European ash trees are showing resistance. Genomic sequencing studies have shown that the whole European population of the fungus was founded from only two genetic strains, illustrating how catastrophic the movement of a microbial species into a new geographical area can be (McMullan et al. 2018).

Human interventions have led to the eradication of a few deadly or economically significant infectious diseases. In 1980, smallpox was eradicated after a long vaccination campaign by the WHO. Rinderpest was the most deadly disease of cattle and was known since the time of the Roman Empire when it spread from Asia into Europe. This disease caused 80–90% of infected cattle to die within 10 days and also resulted in famines for humans. The causative agent is a morbillivirus which also infected over 40 species of wildlife, with devastating effects on African wildlife. Rinderpest was brought gradually under control by cattle vaccination in combination with monitoring of cattle herds and wildlife. An international effort to eradicate the virus was begun in 1994 and was successful with the virus pronounced eradicated on 25th May, 2011 (Morens et al. 2011).

6.2   Localised Severe Natural Events on Earth

Another category of natural risks to life are external risks caused by geological activity, such as earthquakes, volcanic eruptions or tsunamis, or severe weather systems, such as hurricanes, tornados and floods. Events such as wildfires can be initiated by lightning strikes. These natural events can lead to severe local, or more widespread, damage of ecosystems or human-made structures and cause the local death of organisms. However, typically life is not extinguished on a global scale. Moreover, life on earth has adapted by evolution and natural selection to survive and even benefit from these natural, sporadic climate events. For example, high storm winds or flash floods may provide dispersal mechanisms for plant seeds, or may transport birds or seeds to new areas, which they may then colonise. Many plant species that grow in regions where natural wildfires occur actually need fire in order for their seeds to break dormancy and germinate; the Buckthorn family plants of California are an example. The seeds of desert plants are adapted to germinate extremely rapidly after seasonal heavy rains. Thus, these events in themselves do not generally pose unusual risks to life.

Events such as volcanic eruptions are more devastating and result in molten lava flows and ash clouds that render barren the surrounding area, causing large animals to flee and possible extermination of other life forms. The subsequent local desolation provide researchers with opportunities to study how a cleared area become re-colonised by living organisms. This knowledge is relevant to models of how life first colonised the land, or how life might recover from localised or short-term climate change events in future.

During the last 150 years, researchers have studied the consequences of major volcanic eruptions to document how life can return to the affected areas over time. A famous example is the eruption of Krakatau (Krakatau Island, off the coast of Sumatra) during May to August of 1883, after hundreds of years of dormancy (Fig. 6.3). The eruption caused destruction of a major part of the island, total loss of vegetation and local extinction of animal and plant species. Changes to weather due to the ash clouds were experienced thousands of miles away. It is estimated that 36,000 people in surrounding areas lost their lives. A scientific team that visited the island in October 1883 (Verbeek 1884) did not find any surviving plants or animals. The return of species to this island was documented periodically from 1886 onwards.

Fig. 6.3 Photograph of the eruption of Krakatau, 26 August 1883 (*Credit* Library of Congress and https://www.theatlantic.com/magazine/archive/1884/09/the-volcanic-eruption-of-krakatoa/376174/)

Plants returned through dispersal of seeds or spores by sea or wind, and a succession of plant types was noted with ferns, then grasses, then woodland, forming a sequence of environments of differing complexity in the interior (Whittaker et al. 1985). The return of non-marine bird, reptile and animal species has continued for at least a century, with losses as well as gains of species occurring over time (Thornton et al. 1988). Thus, whilst life returned rapidly to the island and the field studies showed the great potential of living organisms to enter a desolate area, the overall balance of terrestrial species on the island was changed by the eruption.

A more recent major volcanic event was the eruption of Mount St. Helens in the Cascades Mountains of the north-western USA. This eruption took place on May 18, 1980 and caused total destruction over a radius of about 8 miles (Fig. 6.4), with the death of 57 people. Trees and animals were affected by the shockwaves over a wider area of about 230 square miles, with the deaths of thousands of land animals and an estimated 12 million salmon. The local landscape was destroyed by fast-moving mudflows from melted snow and glaciers, and air-bourne ash particles from the explosion affected an area of over 2,200 square miles (https://volcanoes.usgs.gov/volcanoes/st_helens/st_helens_geo_hist_99.html).

Fig. 6.4 Destruction of forests on the slopes of Mount St. Helens after the blast caused by the eruption on May 18th, 1980 (Image credit https://www.mshslc.org/gallery/)

In 1982, the USA Congress set aside over 100, 000 acres around the volcano to follow natural changes to the environment over time: this area became the Mount St. Helens National Volcanic Monument. Since the 1980s, Mount St. Helens has continued active with sporadic smaller volcanic eruptions and many earthquakes. The National Volcanic Monument area has provided researchers with an unparalleled opportunity to study how life can return to barren lavafields and ashlands. Amazingly, isolated signs of life were detected even in the first year after the eruption, as burrow-dwelling animals or plants with underground buds that had been protected from the main blast started to re-emerge. The snow cover at the time of the eruption also protected some species. The interactions of surviving organisms with seeds, spores, or organisms blown in by the wind has meant that complex ecosystems started to re-develop relatively rapidly. Microbial populations were key to adding nutrients and moisture-retaining properties to the barren lava flow environments so that plant and animal re-colonisations could start. Twenty-five years later, most species typical of undisturbed forests of the region had been restored (Dale et al. 2005).

Knowledge of how life can recover from a localised severe natural event also gives researchers important clues as to how life on Earth may have recovered from natural catastrophes in the distant past. The fossil record provides evidence for at least five mass extinctions during the history of life (Table 6.1). From the fossil record we can also learn that the species that survive tend give rise to new species by diversifying (radiating) over time. Competition for environmental resources is reduced or altered after the mass extinction event and so different natural selection processes are at work under the new environmental conditions (Hull 2015) (Fig. 6.5). From the available geological and paleo-climatological data, each mass extinction event appears to have had a distinct causation, which may have involved a complex interplay of geological or weather events coupled with cascading effects on food chains once species diversity began to be affected. Massive volcanic activity that brought about changes in the global climate may have been a common factor in several of these natural disasters.

The Ordovician–Silurian and late Devonian mass extinction events are thought to have been caused by global cooling. There is increasing evidence that the highly devasting Permian-Triassic mass extinction was brought about by huge volcanic activity and release of gases in the Siberian region that led to global warming. The average temperatures are thought to have reached 35C– 40C. Increased amounts of atmospheric carbon dioxide and sulphides would also have led to ocean acidification and, along with warming of the oceans, resulted in reduced oxygen saturation of the oceans. The death of marine organisms by oxygen starvation along with the physiological stress of increased temperatures may have been major drivers of this mass extinction (Benton 2018).

Table 6.1 The historical mass extinction events on Earth. Based on Longrich et al. 2011; Hull 2015; Brusatte et al. 2015b; Benton 2018

Extinction Event	Date (MYA)	Severity (estimated % loss of species)	Causation
Ordovician–Silurian	439	86	Glaciation (possibly due to expansion of plants and increased removal of CO_2 from the air); reduced sea levels
Late Devonian	364	70–82	Algal blooms reduced oxygen in the seas; collapse of food chains. Possible Earth cooling due to volcanic ash in the air. May have taken place over thousands of years
Permian–Triassic	251	96	Massive volcanic activity; high CO_2 in air led to increase in methane-producing bacteria; global warming and ocean acidification
Triassic-Jurassic	214–199	76 (*Led to predominance of dinosaursover mammals*)	Asteroid hit? Massive volcanic eruptions leading to climate change
Cretaceous–Paleogene	65	76 (*Extinction of dinosaursapart from avian dinosaurs*)	Asteroid hit at Chicxulub

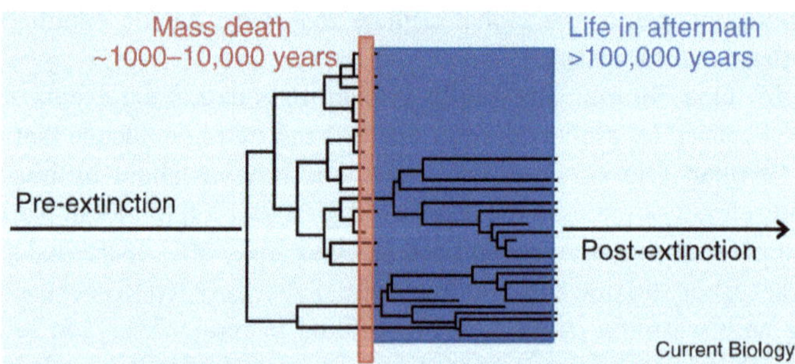

Fig. 6.5 Schematic diagram to illustrate how species diversity and phylogenetic relationships may change due to a mass extinction event. In this hypothetical example, many species die during the extinction event, yet three species survive and diversify genetically over many thousands of years to give rise to many new species. Reproduced with permission from Hull 2015. Elsevier licence number 4780320533891

6.3 Hits by Asteroids and Previous Extinction Events

Astronomical observations of surfaces of other rocky planets and our Moon with increasingly powerful telescopes have revealed large craters caused by collisions with asteroids or comets. Craters also exist on the earth, but have become obscured over time due to the weathering and erosion of rocks that takes place due to Earth's atmosphere. Asteriods and comets are cosmic bodies that originate within our Solar System. Asteroids are mostly formed of rocky material and orbit between Mars and Jupiter, whereas comets are formed at greater distances from the Sun and contain ice and dust from the formation of the Solar System (https://cneos.jpl.nasa.gov/about/basics.html). The same object is termed an asteroid when it travels through cosmic space, a meteor when the fragment vaporises and burns up with a flash in the Earth's atmosphere, and a "meteorite" if it lands on Earth. Meteoroids are fragments that have broken off from an asteroid or a comet. For our ancestors, meteors and comets appeared suddenly in the sky and their origin was unknown. They were therefore often regarded as portents or religious miracles.

Asteroid bombardment of the Earth was more active in the early history of our Solar system, up to about 3.6 billion years ago, and the formation of our Moon resulted from the impact of a huge body with the original earth (Jacobson et al. 2014). When impacts such as these take place, high energy and heat are transferred to the receiving body and cause rocks to become vaporised. The resulting droplets then cool in the Earth's atmosphere and become deposited back on the ground as rock globules termed spherules. The detection of spherule layers by geologists therefore provides insights into the timings of cosmic bombardment (Johnson and Melosh 2012). It has been proposed that repeated asteroid hits, causing melting and remodelling of the rocks of the early Earth 4.5–4 billion years ago, may explain why very little of the geological record of Earth before about 3.8 billion years ago can be identified by geological explorations (Marchi et al. 2014) (also discussed in Chap. 3).

In July 1994, millions of people were able to watch, via telescopes connected to television, a huge encounter of fragments from a comet named Shoemaker-Levy with the planet Jupiter. This comet, estimated to be about 1.5 km–2 km in diameter, had fragmented under the influence of Jupiter's gravitational field in 1992. After the collisions in 1994, large scars (bigger than the entire Earth) became visible at the surface of Jupiter (https://solarsystem.nasa.gov/asteroids-comets-and-meteors/comets/p-shoemaker-levy-9/in-depth/). On the Earth, the occurrence of a similar event in the present day would probably extinguish a large fraction of living organisms. Huge dust clouds produced by the collision of huge masses driven would reduce the sunlight reaching the Earth to near-zero levels. Depending on how long this dust lasted, species dependent on photosynthesis and therefore energy from the Sun's irradiation would likely die and over time become extinct. The loss of these plants and bacteria would then have consequences for all the living organisms that depend on them in the food chain.

Studies of the Earth have identified around 100 craters, some resulting from hits by bodies at least 5 km in diameter. Although it appears from geological evidence

(Table 6.1) that some of the historic mass extinctions were caused by large-scale volcanic activity, asteroid hits have also been discussed as possible causes of mass extinctions (Rampino et al. 1997). For the mass extinction that took place at the end of the Cretaceous period, around 65.5 million years ago, there is now good evidence that the precipitating event was the collision of a large (estimated to be at least 11 km in diameter) asteroid with the Earth. The Cretaceous–Paleogene extinction event is best known for the mass extinction of non-avian dinosaurs and ammonites, but in fact around 75% of all life on Earth was lost (Schulte et al. 2010). The asteroid hit the Earth at Chicxulub in the Yucatan peninsula, on the Gulf of Mexico. The impact site remains marked by a crater about 200 km in diameter and a geological layer of ejecta (fragmented material from the impact crater) that can be identified across the world (Schulte et al. 2010). It is estimated that around 425 Gigatonnes of carbon dioxide and 325 Gigatonnes of sulphur were released by the impact (Gulick et al. 2019).

The geological record within the crater has been investigated by drilling down 750 m at a key location. The materials and layers identified demonstrate how the Cretaceous period ended so dramatically, with over 100 m of impact-melted rock covered by metres of impact fragments that were probably deposited by flowing, molten rocks. The overlying deposited materials provide evidence for a tsunami wave that propagated from the crater to the ocean, followed by earthquakes and turbulent resurges of water. A layer of charcoal attests to widespread forest fires. The fires are thought to have been set alight by the impact itself or by the ejected heated rock fragments. These turbulent and life-destroying events marked the beginning of the Cenozoic era on Earth. As a result of the large-scale fires, the Earth's atmosphere continued to contain high levels of particulate material during the following years (Gulick et al. 2019), which led to cooling of the Earth's surface. It has been modelled that the sunlight reaching the Earth's surface could have been decreased by as much as 20%. The fossil record of plankton and the presence in sediments of lipids produced by crenarchaeota plankton also provide empirical evidence that a rapid decrease in surface sea temperatures took place over months to decades (Vellekoop et al. 2014).

Another consequence of the asteroid hit was a massive release of carbon dioxide into the atmosphere after the impact. This led to rapid ocean acidification and, in turn, to loss of marine calcifying organisms in the plankton. For example, many species of foraminifera (a type of single-celled marine plankton with an external calcium carbonate shell) went extinct and ocean food-chains collapsed (Schulte et al. 2010). It is estimated from combined studies of fossil plankton and geochemical analyses that ocean ecosystems took around 30,000 years to recover after the asteroid hit (Lowery et al. 2018). However, other studies that have tracked nanoplankton indicate a much longer timeline, with populations of the species present after the extinction event remaining very changeable and vulnerable to further environmental changes over the next 1.8 million years (Alvarez et al. 2019).

There is much evidence that the Chicxulub asteroid hit had dramatic consequences for land plants and animals in addition to the dinosaurs. After the initial high losses of species (as shown by the dramatic changes in the fossil record), the first recovery of plants on land involved spore-forming ferns and fungi, followed by a rise of angiosperms (flowering plants) over coniferous plants, leading overall to a different range of plant species on land (Vajda and Bercovici 2014). In parallel with the extinction of non-avian dinosaurs and these changes in vegetation, the surviving mammal species diversified into new ecological niches that had been previously dominated by dinosaurs (Novacek 1999). Birds (which had evolved from therapod dinosaurs by the time of the early Cretaceous) also underwent a partial extinction followed by re-diversification, such that representatives of modern birds are known in the fossil record only from the Paleocene and Eocene onwards (Brusatte et al. 2015a, Longrich et al. 2011). A study of a remarkable fossil site in Colorado has revealed how the recovery and origination of diverse types of land plants may have enabled the evolution of larger land mammals, and has emphasised how the recovery of ecosystems took place over hundreds of thousands of years (Lyson et al. 2019).

An inherent limitation of the fossil record is that it provides a partial view on life on Earth in the deep past. However, modern large-scale methods such as network analysis can be applied to fossil datasets to give a more integrated view on how life has changed over time. For example, a recent network study of marine animal fossils from the Paleobiology Database searched for instances of co-occurrence of fossils previously identified according to taxonomic groupings. The concept was to identify which species consistently co-occurred and were therefore likely to have been part of the same ecological community. The change in communities over time was then examined to gain a ecosystem-scale view of mass extinction events. As an example, Figure 6.6 presents three networks of species (red, yellow and blue), identified according to criteria of taxonomical orders, that have persisted over time and been affected in different ways by mass extinction events. The "red" community

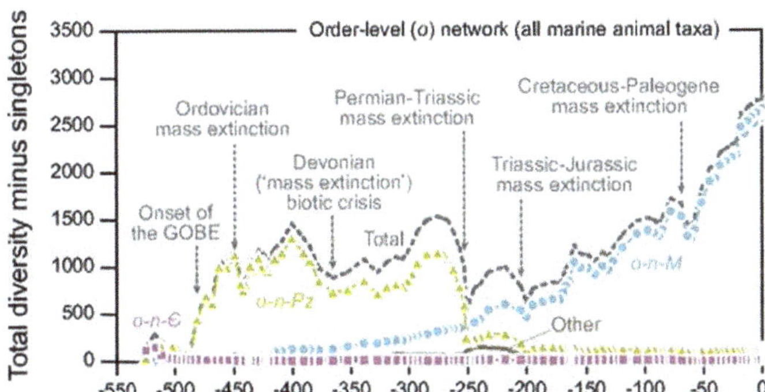

Fig. 6.6 Plot of mass extinction events over time based on a network analysis of marine animal fossils. The X axis shows millions of years before the present. The different coloured traces each represent the abundance of fossils (categorised here by taxonomical order) that were identfied by network analysis to be part of the same community. The plot shows how the relative diverity of these communities has changed over time. The "yellow" community was the most diverse before the Permian-Triassic mass extinction and since that event the "blue" community has become the most diverse. GOBE = Great Ordovician Biodiversification Event. From Muscente et al. 2018 under CC BY-NC-ND 4.0

emerged early and has remained at low diversity ever since. The "yellow" community also emerged early, and rapidly increased in diversity but was very strongly affected by the Permian-Triassic mass extinction, such that the diversity of this community has never since recovered. The later emerging 'blue" community became the most diverse community after the Permian–Triassic extinction and has continued to increase in diversity even after two more recent mass extinction events. This type of analysis provides a new approach to assess in a broad way how ecological communities were affected by the historic mass extinctions (Fig. 6.6) (Muscente et al. 2018).

To obtain a direct feeling of what happens when an extra-terrestrial body enters the atmosphere of the Earth, we can consider the rather detailed recordings of an airburst that happened over the Chelyabinsk Oblast in Kazakhstan on February 15, 2013 (Popova et al. 2013). This was the second disaster caused by an asteroid that was well-recorded by humans: the first occurred over Tunguska, Siberia, Russia, in 1908 and was probably also caused by an asteroid airburst.

Estimates of the Russian Federal Space Agency indicate that the asteroid of 2013 was about 20 m in diameter with a weight of 12,000 tons. It travelled at a speed of 30 km/s in a low trajectory to the Earth's surface. When it reached the atmosphere its speed was reduced to about 15 km/s due to friction with the gases of the atmosphere. Precise measurements were possible because the path and speed of the meteor was monitored by a large number of military stations established by the Russians. At about 30 km above the earth's surface, the meteor exploded (the airburst) generating an extremely bright flash (Fig. 6.7). A cloud of hot dust and gases was produced and the meteor fragmented into many pieces. When these meteorite fragments hit the ground they caused an earthquake of low strength.

The Chelyabinsk meteor blast caused extensive damage over an area 100 km wide and several tens of kilometres long. Several hundred people were injured, mainly from

Fig. 6.7. Photograph of the meteor that entered the atmosphere of the Earth near the town of Chelyabinsk in Kasachstan in February 2013 (Reproduced from Wikipedia, English version)

collapsing buildings. Flash blindness and ultraviolet burns were reported, yet, fortunately, nobody was killed. Chelyabinsk is a medium-sized town and the surrounding area is sparely populated. The total energy generated in the blast was estimated to be equivalent to 400–500 kilotons of TNT. This is about 2.5 times the energy released from the atomic bomb detonated at Hiroshima (see Chap. 7) and less than that released by modern hydrogen bombs. Nevertheless, because most of the energy of meteor explosions is dispersed into the atmosphere and there is no radioactive fallout, asteroids of this size are less dangerous to life than nuclear bomb explosions (see Chapter 7 for a discussion of the risks of nuclear warfare).

Encounters of the Earth with asteroids are very difficult to prevent. Ballistic interference with asteroid pathways was proposed shortly after the disaster in Kazakhstan. However, because of the high costs involved, shooting at asteroids has not been attempted. Forecasts and warnings of asteroid trajectories are possible because of continuous observations of the sky, use of satellites, and precise calculations of the pathways of known asteroids and newly detected bodies. NASA and other agencies enforced detection programmes after the Chelyabinsk catastrophe, and many small asteroids are continuously detected (Fig. 6.8). Potentially hazardous objects are classified as those with trajectories that would bring them within ~ 7,480,000 km of Earth, and particular attention is given to objects larger than 140 m in diameter because of their high potential to cause widespread damage. NASA's Sentry system monitors the motion of asteroids for possible collisions with Earth within the next 100 years. The large asteroid JF1, first detected in 2009, is one example. From the current data, NASA considers that there is a 0.026% chance of this asteroid hitting the Earth in 2022 (https://cneos.jpl.nasa.gov/sentry/).

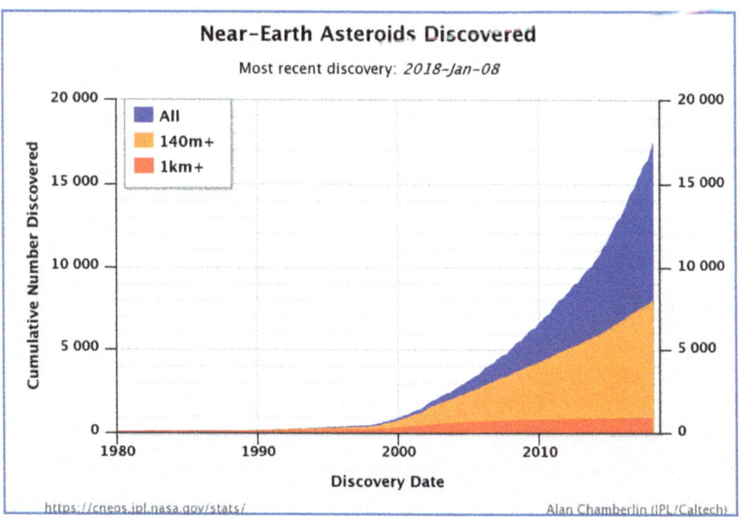

Fig. 6.8 Number of near Earth asteroids of different diameters detected up to 2018, as monitored by NASA (Reproduced with permission from http://nasa.gov.stats/)

It is undisputed that serious natural catastrophes occurred on Earth during the last 4 billion years, some resulting from asteroid hits. A key question for the topic of this book is whether cosmic events such as asteroid hits could abolish life on the modern Earth. As the studies of the Chicxulub asteroid hit have shown, a hit from a large body led to massive changes to the species and communities living on the Earth, and it may take thousands, if not millions, of years for the diversity of life to recover and a new steady-state to be reached. Thus, an asteroid hit re-directed biological evolution. Theoretical models have also been constructed, based on data from the fossil record over time and predictions from computational models. It appears that mass extinction events have an extremely low probability, although it is very difficult to assign an accurate probability to very rare events. The probability of the Earth being hit by a large meteor travelling from another star system also appears very low, although the detection of these inter-stellar bodies began only in 2017 (Meech et al. 2017). The study of Chicxulub indicates that an asteroid hit from 140 m diameter body would cause massive devastation but would probably not abolish all life. Evolution would continue and life would re-diversify over thousands to millions of years.

References

Agriculture Victoria (2017) Boosting rabbit biocontrol: RHDV1 K5 in Victoria. https://agriculture. vic.gov.au/agriculture/pests-diseases-and-weeds/pest-animals/invasive-animal-management/est ablished-invasive-animals/rhd-boost

Alvarez SA, Gibbs SJ, Bown PR, Kim H, Sheward RM, Ridgwell A (2019) Diversity decoupled from ecosystem function and resilience during mass extinction recovery. Nat 574:242–245

Basu A, Zinger T, Inglima K, Woo KM, Atie O, Yurasits L, See B, Aguero-Rosenfeld ME (2020) Performance of Abbott ID NOW COVID-19 rapid nucleic acid amplification test in nasopharyn-geal swabs transported in viral media and dry nasal swabs, in a New York City academic institution. J Clin Microbiol May 29 2020:JCM.01136–20. https://doi.org/10.1128/JCM.01136-20

Benton MJ (2018) Hyperthermal-driven mass extinctions: killing models during the Permian-Triassic mass extinction. Philos Trans R Soc A. https://doi.org/10.1098/rsta.2017.0076

Berche P (2012) Louis Pasteur, from crystals of life to vaccination. Clin Microbiol Infect 18(Suppl 5):1–6. https://doi.org/10.1111/j.1469-0691.2012.03945.x

Brusatte SL, Jingmai K O'Connor, Erich D Jarvis (2015a) The origin and diversification of birds. Curr Biol 25(19):R888–98

Brusatte SL, Butler RJ, Barrett PM, Carrano MT, Evans DC, Lloyd GT, Mannion PD, Norell MA, Peppe DJ, Upchurch P, Williamson TE (2015b) The extinction of the dinosaurs. Biol Rev Camb Philos Soc. 90:628–642. https://doi.org/10.1111/brv.12128

Dale VH, Crisafulli CM, Swanson FJ (2005) 25 Years of ecological change at Mount St. Helens. Sci 308:961–962. https://doi.org/10.1126/science.1109684

Domingo JL, Marquès M, Rovira J (2020) Influence of airborne transmission of SARS-CoV-2 on COVID-19 pandemic. A review Environ Res 188:109861. https://doi.org/10.1016/j.envres.2020. 109861

Duffy S (2018). Why are RNA virus mutation rates so damn high? PLoS Biology 16(8):e3000003. https://doi.org/10.1371/journal.pbio.3000003

Ellis H. (2020). Florence Nightingale: creator of modern nursing and public health pioneer. J Perioper Pract 30:145–146. https://doi.org/10.1177/1750458919851942

Glansdorff P, Prigogine I (1971) Thermodynamics theory of structure, stability and fluctuations. Wiley-Interscience

Gulick SPS, Bralower TJ, Ormö J, Hall B, Grice K, Schaefer B, Lyons S, Freeman KH, Morgan JV, Artemieva N, Kaskes P, de Graaff SJ, Whalen MT, Collins GS, Tikoo SM, Verhagen C, Christeson GL, Claeys P, Coolen MJL, Goderis S, Goto K, Grieve RAF, McCall N, Osinski GR, Rae ASP, Riller U, Smit J, Vajda V, Wittmann A, Expedition 364 Scientists (2019) The first day of the Cenozoic. Proc Natl Acad Sci USA 116:19342–19351. https://doi.org/10.1073/pnas.190 9479116

Hadfield J, Megill C, Bell SM, Huddleston J, Potter B, Callender C, Sagulenko P, Bedford T, Neher RA (2018) Nextstrain: real-time tracking of pathogen evolution. Bioinf 34(23):4121–4123. https://doi.org/10.1093/bioinformatics/bty407

Hadjadj J et al (2020) Impaired type I interferon activity and inflammatory responses in severe COVID-19 patients. Sci 369:718–724

Hall RN, Capucci L, Matthaei M, Esposito S, Kerr PJ, Frese M, Strive T (2017) An in vivo system for directed experimental evolution of rabbit haemorrhagic disease virus. PLoS One 12(3):e0173727

Hull P (2015) Life in the aftermath of mass extinctions. Curr Biol 25:R941–R952

Jacobson SA, Morbidelli A, Raymond SN, O'Brien DP, Walsh KJ, Rubie DC (2014) Highly siderophile elements in Earth's mantle as a clock for the moon-forming impact. Nat 508:84–87. https://doi.org/10.1038/nature13172pmid:24695310

Jamrozi E, Little K, Bulla S, Emerson C, Kang G, Kapulugh M, Reyio E, Saenz C. Shah S, Smith PG, Upshur R, Weijern C, Selgelid MJ for the WHO Working Group for Guidance on Human Challenge Studies in COVID-19. 2021. Key criteria for the ethical acceptability of COVID-19 human challenge studies: Report of a WHO Working Group. Vaccine 39, 633–640.

Johnson BC, Melosh HJ (2012) Impact spherules as a record of an ancient heavy bombardment of Earth. Nat 485:75–77. https://doi.org/10.1038/nature10982

Krupovic M, Dolja VV, Koonin EV (2019) Origin of viruses: primordial replicators recruiting capsids from hosts. Nat Rev Microbiol 17:449–458. https://doi.org/10.1038/s41579-019-0205-6

Lederberg J (1997) Infectious disease as an evolutionary paradigm. Emerg Infect Dis 3(4):417–423. https://doi.org/10.3201/eid0304.970402

Longrich NR, Tokaryk T, Field DJ (2011) Mass extinction of birds at the Cretaceous-Paleogene (K–Pg) boundary. PNAS 108:15253–15257

Louca S, Mazel F, Doebeli M, Parfrey LW (2019) A census-based estimate of Earth's bacterial and archaeal diversity. PLoS Biol 17(2):e3000106. https://doi.org/10.1371/journal.pbio.3000106

Lowery CM, Bralower TJ, Owens JD3, Rodríguez-Tovar FJ4, Jones H2, Smit J5, Whalen MT6, Claeys P7, Farley K8, Gulick SPS9, Morgan JV10, Green S11, Chenot E12, Christeson GL9, Cockell CS13, Coolen MJL14, Ferrière L15, Gebhardt C16, Goto K17, Kring DA18, Lofi J19, Ocampo-Torres R20, Perez-Cruz L21, Pickersgill AE22,23, Poelchau MH24, Rae ASP10, Rasmussen C9, Rebolledo-Vieyra M25, Riller U26, Sato H27, Tikoo SM28, Tomioka N29, Urrutia-Fucugauchi J21, Vellekoop J7, Wittmann A30, Xiao L31, Yamaguchi KE32,33, Zylberman W34 (2018) Rapid recovery of life at ground zero of the end-Cretaceous mass extinction. Nat 558(7709):288–291. https://doi.org/10.1038/s41586-018-0163-6

Lyson TR, Miller IM, Bercovici AD, Weissenburger K, Fuentes AJ, Clyde WC, Hagadorn JW, Butrim MJ, Johnson KR, Fleming RF, Barclay RS, Maccracken SA, Lloyd B, Wilson GP, Krause DW, Chester SGB (2019) Exceptional continental record of biotic recovery after the Cretaceous-Paleogene mass extinction. Sci 366:977–983. https://doi.org/10.1126/science.aay2268

McMullan M, Rafiqi M, Kaithakottil G, Clavijo BJ, Bilham L, Orton E, Percival-Alwyn L, Ward BJ, Edwards A, Saunders DGO, Garcia Accinelli G, Wright J, Verweij W, Koutsovoulos G, Yoshida K, Hosoya T, Williamson L, Jennings P, Ioos R, Husson C, Hietala AM, Vivian-Smith A, Solheim H, MaClean D, Fosker C, Hall N, Brown JKM, Swarbreck D, Blaxter M, Downie JA, Clark MD (2018) The ash dieback invasion of Europe was founded by two genetically divergent individuals. Nat Ecol Evol 2:1000–1008. https://doi.org/10.1038/s41559-018-0548-9

Marchi S, Bottke WF, Elkins-Tanton LT, Bierhaus M, Wuennemann K, Morbidelli A, Kring DA (2014) Widespread mixing and burial of Earth's Hadean crust by asteroid impacts. Nat 511:578–582. https://doi.org/10.1038/nature13539

Meech KJ, Weryk R, Micheli M, Kleyna JT, Hainaut OR, Jedicke R, Wainscoat RJ, Chambers KC, Keane JV, Petric A, Denneau L, Magnier E, Berger T, Huber ME, Flewelling H, Waters C, Schunova-Lilly E, Chastel S (2017) A brief visit from a red and extremely elongated interstellar asteroid. Nat 552(7685):378–381

Morens DM, Holmes EC, Davis AS, Taubenberger JK (2011) Global rinderpest eradication: lessons learned and why humans should celebrate too. J Infect Dis 204:502–505. https://doi.org/10.1093/infdis/jir327

Muscente AD, Prabhu A, Zhong H, Eleish A, Meyer MB, Fox P, Hazen RM, Knoll AH. 2018. Quantifying ecological impacts of mass extinctions with network analysis of fossil communities. Proc Natl Acad Sci U S A 115(20):5217–5222. https://doi.org/10.1073/pnas.1719976115.

Mushegian AR (2020) Are there 10^{31} virus particles on Earth, or more, or fewer? J Bacteriol 202:e00052–e120. https://doi.org/10.1128/JB.00052-20

Nature Reviews Microbiology Editorial (2011) Microbiology by numbers. Nat Rev Microbiol 9:628. https://doi.org/10.1038/nrmicro2644

Novacek M (1999) 100 million years of land vertebrate evolution: the Cretaceous-early Tertiary transition. Ann Mo Bot Gard 86:230–258

Pairo-Castineira E et al (2020) Genetic mechanisms of critical illness in Covid-19. https://doi.org/10.1101/2020.09.24.20200048

Popova, OP, Jenniskens P, Vacheslav E, Kartashova A, Biryukov E, Khaibrakhmanov S, Shuvalov V, Rybnov Y, Dudorov A, Grokhovsky VI, Badyukov, Dmitry D, Yin, Q-Z, Gural PS; Albers J, Granvik M, Evers, Läslo G, Kuiper J, Kharlamov V, Solovyov A, Rusakov YS, Korotkiy S, Serdyuk I, Korochantsev AV, Larionov MY, Glazachev D, Mayer AE, Gisler G, Gladkovsky SV, Wimpenny J, Sanborn ME, Yamakawa A, Verosub KL, Rowland DJ, Roeske S, Botto NW, Friedrich JM, Zolensky ME, Le L, Ross D, Ziegler, Nakamura T, Ahn I; Lee JL, Zhou Q, Li X-H, Li Q-L, Liu Y, Tang G-Q, Hiroi T, Sears D, Weinstein IA, Vokhmintsev AS, Ishchenko AV, Schmitt-Kopplin P, Hertkorn N, Nagao K, Haba MK, Komatsu M, Takashi M (the Chelyabinsk Airburst Consortium) (2013). Chelyabinsk airburst, damage assessment, meteorite recovery, and characterization. Science 342:1069–1073

Potter C, Harwood T, Knight J, Tomlinson I (2011) Learning from history, predicting the future: the UK Dutch Elm disease outbreak in relation to contemporary tree disease threats. Philos Trans R Soc Lond B Biol Sci 366:1966–1974. https://doi.org/10.1098/rstb.2010.0395

Powell WA, Newhouse AE, Coffey V (2019) Developing Blight-Tolerant American chestnut trees. Cold Spring Harb Perspect Biol 11(7):a034587. https://doi.org/10.1101/cshperspect.a034587

Rampino MR, Haggerty BM, Pagano TC (1997) A unified theory of impact crises and mass extinctions: quantitative tests. Ann N Y Acad Sci 822:403–431

Riedel S (2005) Edward Jenner and the history of smallpox and vaccination. Proc (bayl Univ Med Cent) 18(1):21–25. https://doi.org/10.1080/08998280.2005.11928028

Schulte P, Alegret L, Arenillas I, Arz JA, Barton PJ, Bown PR, Bralower TJ, Christeson GL, Claeys P, Cockell CS, Collins GS, Deutsch A, Goldin TJ, Goto K, Grajales-Nishimura JM, Grieve RA, Gulick SP, Johnson KR, Kiessling W, Koeberl C, Kring DA, MacLeod KG, Matsui T, Melosh J, Montanari A, Morgan JV, Neal CR, Nichols DJ, Norris RD, Pierazzo E, Ravizza G, Rebolledo-Vieyra M, Reimold WU, Robin E, Salge T, Speijer RP, Sweet AR, Urrutia-Fucugauchi J, Vajda V, Whalen MT, Willumsen PS (2010) The Chicxulub asteroid impact and mass extinction at the Cretaceous-Paleogene boundary. Sci 327:1214–1218. https://doi.org/10.1126/science.1177265

Smith DA (2000) American chestnut: ill-fated monarch of the eastern hardwood forest. J Forest 98:12–15. https://doi.org/10.1093/jof/98.2.12

Smith RA, Nkuchia MM, Andrew FR (2015). Antibiotic resistance: a primer and call to action. Health Commun 30: 309–314. https://doi.org/10.1080/10410236.2014.943634

Spiesschaert B, McFadden G, Hermans K, Nauwynck H, Van de Walle GR (2011) The current status and future directions of myxoma virus, a master in immune evasion. Vet Res 42:76. https://doi.org/10.1186/1297-9716-42-76

Suzuki TA, Ley RE (2020) The role of the microbiota in human genetic adaptation. Sci 370:1180. https://doi.org/10/1126.science.aaz6827

Thornton IW, Zann RA, Rawlinson PA, Tidemann CR, Adikerana AS, Widjoya AH (1988) Colonization of the Krakatau islands by vertebrates: equilibrium, succession, and possible delayed extinction. Proc Natl Acad Sci USA 85(2):515–518

Vajda V, Bercovici A (2014) The global vegetation pattern across the Cretaceous-Paleogene mass extinction interval: a template for other extinction events. Glob Planet Change 122:29–49

Vellekoop J, Sluijs A, Smit J, Schouten S, Weijers JW, Sinninghe Damsté JS, Brinkhuis H (2014) Rapid short-term cooling following the Chicxulub impact at the Cretaceous-Paleogene boundary. Proc Natl Acad Sci U S A 111, 7537–41. https://doi.org/10.1073/pnas.1319253111

Verbeek RDM (1884) The Krakatoa eruption. Nat 30:10–15

Whittaker, Tagawa H, Suzuki E, Partomihardjo T, Suriadarma A (1985) Vegetation and succession on the Krakatau islands, Indonesia.Veg 60:151–145

Zeberg H, Pääbo S (2020) The major genetic risk factor for severe COVID-19 is inherited from Neanderthals. Nat. https://doi.org/10.1038/s41586-020-2818-3

Zhou P, Yang X, Wang X et al (2020) A pneumonia outbreak associated with a new coronavirus of probable bat origin. Nat 579:270–273. https://doi.org/10.1038/s41586-020-2012-7

Chapter 7
Human-Made Risks from Nuclear and Chemical Warfare

In Chap. 6 we discussed the natural web of life and predator-prey relationships. For humans, another threat to life emerges from the many wars between different groups of people. Fighting and wars appear to have featured from the very beginning of human history. The reasons for war can be found in a strong competition and aggressiveness between different groups, which appears to be intrinsic to the character of humans, along with the intelligence and practical skills to devise new tactics and technologies for warfare. In modern times, the decision to make war comes from political leaders who are often guided by their own interests and aggressions. Jean Paul F. Richter (1854) was a German poet. He suggested an improvement for war prevention: *"Up to now the misfortune of the earth was that two persons decided on war but millions had to fight and suffer from it. It would be better, but also not perfect, that millions would decide and two would fight"*. Jean Paul also defined the reasons for a war: *"Each sin houses the war similar to a spark and a burst of fire"*.

In the twentieth century, risks from human actions grew to ever more dangerous levels. This was due in part to the explosive growth of the global human population and also to conflicts between nations and ideological groups that were supported by rapid growth in military technologies. In the first World War, it is estimated that 9 million soldiers and 10 million civilians were killed and Europe recovered only slowly. The second World War was even more cruel and disastrous, with very widespread deaths in civilian populations from attacks involving new weapons, aerial bombardments, the first nuclear weapons, and targeted genocide by Nazi-Germany. Worldwide, it is estimated that between 50 million and 85 million people were killed.

Today we speak of wars waged with conventional weapons and wars with nuclear and other modern weapons of mass destruction. A nuclear explosion creates deadly radiation and radioactive fallout that render life impossible over a large area and which also spread widely in the atmosphere. This means that dangerous radioactive compounds become widely distributed and, according to their radioactive half-life, persist for long periods of time. Many of these compounds become incorporated into living organisms through the food chain, leading to increased risks of cancer in humans. From the invention of nuclear weapons, it has been recognised that their use

J. C. Adams and J. Engel, *Life and Its Future*,
https://doi.org/10.1007/978-3-030-59075-8_7

poses very major risks, and has the potential to destroy all life on earth, or turn biological evolution in a new direction. A second devastating weapon of mass destruction, in use since the First World War, is the chemical weapon. Nowadays, weapons of mass destruction are planned and manufactured in most countries. Politically, they are justified as defence weapons or even as peace-keeping weapons. In political speeches it is often stated that peace is guaranteed by the deterrence produced by many thousands of nuclear weapons and tons of poisonous gases.

In this chapter, we discuss the use of nuclear and chemical agents in warfare, current treaties and restrictions, and the threats that are posed to all living organisms as well as to humans. The most dangerous risks come from activities in which high state technology could be purposefully applied to destroy life. The truth is that modern nuclear weapons render it possible to destroy the biological world.

7.1   Nuclear Weapons

7.1.1   Beginning of the Era of Nuclear War

In August 1945 the war against Nazi-Germany had finished but the war with Japan continued. The official reason for the invention of nuclear bombs was to end this war. The era of nuclear war was initiated by the complete destruction, by two atomic fission bombs, of Hiroshima on August 6, 1945 and of Nagasaki three days later (Fig. 7.1). At this time JE lived in Germany, far away from Japan. He had seen terrible destruction of cities during the Second World War, but had never seen pictures of cities which were completely razed to the ground like Hiroshima and Nagasaki (Fig. 7.2).

Fig. 7.1 Atomic bomb mushroom clouds over Hiroshima on August 6, 1945 (left), and Nagasaki on August 9, 1945 (right) (Reproduced from Wikipedia, English version 2020)

Fig. 7.2 Hiroshima after the detonation of "Little boy" (Reproduced from Wikipedia, English version 2020)

Later he learned in school that each of the atomic bombs had released energy equivalent to 100–200 kilotons of trinitrotoluene (TNT). The energy of the explosion of TNT is used as a standard for the energies of explosions. A ton of TNT is equivalent to $4.18\ 10^9$ joules or 1.162×10^3 kilowatt hours. This enormous energy was contained in a bomb containing about 1 kg of uranium or plutonium. The Hiroshima bomb was named "Little boy" by its inventors and the Nagasaki bomb was named "Fat boy". The bombings indeed ended the war with Japan immediately and Japan was occupied by the USA.

It is estimated that about 70,000 people were killed immediately during the bombing of Hiroshima due to acute radiation poisoning and burns caused by the intense heat and that 90% of Hiroshima city was destroyed. Within 1 km of the blast centre, 50% of the population are thought to have died. The number of deaths continued to increase over time due to radiation-related sickness, injuries and a high rate of cancers, and is estimated to have reached around 300,000. A first report, based on the accounts of six survivors, was published by the journalist John Hersey in the New Yorker magazine (Hersey 1946a) and then as a book (Hersey 1946b). It has been an open discussion whether or not the step into nuclear war was necessary to end the second World War. At the time of the nuclear bombings, the war was almost won by the Allies. It has been discussed whether a decisive outcome could perhaps have been achieved by conventional weapons. An alternative reason for the decision to use the nuclear weapons may have been the growing tensions between the USA and the Soviet Union. At the 70th anniversary commemorations in Hiroshima in 2015, the Mayor, Kasumi Matsui, said "I don't think we should make an issue of whether they (the USA) should apologise or not… What I want leaders to do when they visit Hiroshima is to vow toward the future that they will never allow this kind of thing to happen again" (reported August 6, 2015, www.thejournal.ie).

7.1.2　The Developers of Nuclear Weapons: J. Robert Oppenheimer

The development of nuclear weapons has a long history and is part of the field of nuclear physics. It is interesting to consider the feelings of scientists who were leading theoretical nuclear physicists at top American universities in the mid-twentieth century and who became leading figures in the development of the first atomic bombs. An example is J. Robert Oppenheimer, who had a decisive influence in the development of the first bomb. He was much against the second bomb on Nagasaki, because he doubted its military necessity. After the first test explosion of a bomb he commented according to Pontin (2007) about the decision to drop the bomb:

> We knew the world would not be the same. A few people laughed, a few people cried, most people were silent. I remembered the line from the Hindu scripture the Bhagavad Gita. Vishnu is trying to persuade the prince that he should do his duty and to impress him takes on his multiarmed form and says, "Now, I am become Death, the destroyer of worlds." I suppose we all thought that one way or another.

After the war, Oppenheimer became an advisor to the Atomic Energy Commission and argued strongly for a control of nuclear weapons. In 1954, he was called to a tribunal of the Atomic Energy Commission to answer questions on his possible past communist involvement and whether he was acting as a spy for the Soviet Union. Edward Teller, a nuclear physicist whose research was in the area of nuclear fusion weapons, testified against Oppenheimer during the hearing. As a result of the hearing, Oppenheimer lost his national security clearance and retired from a public role. These events caused anger in the international community of physicists (https://www.ato micheritage.org/history/oppenheimer-security-hearing).

7.1.3　The Potential of Today's Nuclear Weapons

The historical atomic fission bombs "Little boy" and "Fat boy" have now been replaced by fusion bombs, usually called hydrogen bombs. Fission bombs operated by the splitting of uranium or plutonium atoms into elements of lower molecular mass. This is the type of reaction that runs in conventional nuclear power plants. In hydrogen bombs, the energy production originates from the fusion of hydrogen atoms into helium, the same process which powers the sun. The energy liberated by hydrogen bombs is much higher than that of fission bombs.

The historical atomic fission bombs were brought to their targets by oversize bombers. Nowadays, the standard carriers are missiles. By far the largest numbers of weapons are held by the USA and Russia, and during the Cold War (the period of rivalry for supremacy between the USA and Soviet Union that began after World War II), there were about 70,000 nuclear weapons in the world. Over time, the USA and the Soviet Union entered into bi-lateral agreements that set limits on the numbers and types of weapons they would hold (discussed further in Sect. 7.1.4). The present long-distance types of missiles have the potential to carry nuclear warheads with an energy equivalent to 20 Megatons TNT from the USA to Russia and vice versa. Less powerful, short-range missiles are stationed along the borders of military blocks. Small nuclear heads can be carried on submarines, as torpedoes. Preparations also include sophisticated early warning systems to detect flying missiles. Several types of nuclear warheads exist. The most dangerous are the so-called dirty heads, which expel radioactive compounds that mostly have long half-lives. We live in a split society: handling of all radioactive compounds in research laboratories is tightly controlled to avoid health risks, whereas radioactive compounds are also used extensively in weapons produced for mass destruction.

All these preparations cost extremely large amounts of money. According to a 2019 estimate, there are around 14,000 nuclear warheads in the world. Figure 7.3 presents an overview of the distribution of nuclear weapons. Through NATO, the USA has hundreds of nuclear weapons stationed in several European countries (https://www. nti.org/analysis/articles/nato-nuclear-disarmament/). The numbers presented should be taken as approximations only, because of the likelihood of additional warheads which are hidden by military secrets.

Nowadays, hydrogen bombs with higher energies can be built by designing multi-step bombs. However, limitations on bomb size are enforced by the limited carrying power of missiles. Russian president Vladimir Putin revealed several new classes of nuclear weapons in his state-of-the-union speech in July 2018 and showed videos of them. One of them has been named Skyfall SSC-X-9 by NATO. It is a nuclear-powered cruise missile that is launched into the air and takes a weaving path at low altitudes. In contrast to normal missiles, this path is not predictable by ballistic geometry, meaning that this missile is virtually undetectable by anti-missile systems. The

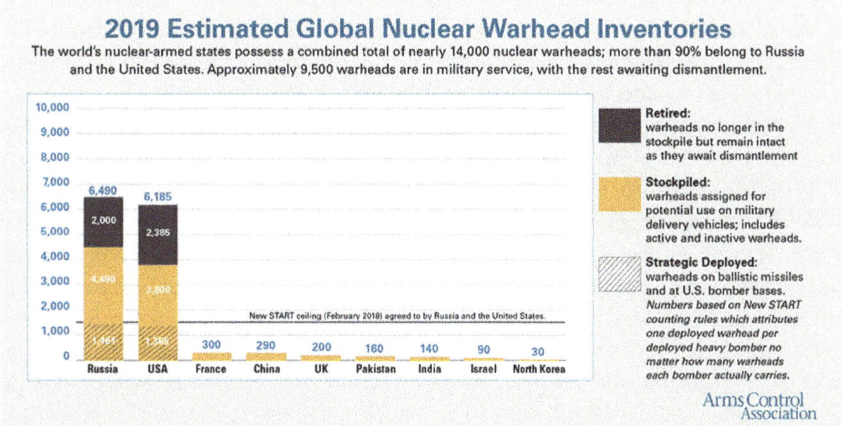

Fig. 7.3 Estimated stocks of nuclear weapons in 2019. Key: Black, weapons retired but not yet dismantled; orange, stockpiled; cross-hatched, on ballistic missiles or deployed at bomber bases (*Source* Hans M. Kristensen and Matt Korda, U.S. Department of State; Stockholm International Peace Research Institute, 2019. Adapted from armscontrol.org)

nuclear-powered cruise missiles can travel very long distances, if necessary several times around the world. Missiles equipped with several dirty nuclear heads are particularly dangerous as they may contaminate large areas (up to whole continents) with dangerous, long-lasting radioactive compounds. This possibility was mentioned by a Russian military representative on Swiss Television and indicates that a step of utmost danger for life on earth is potentially in reach. Details about SSC-X-9 are covered by state secrets; however, a recent nuclear explosion of the size of the Chernobyl nuclear plant explosion in a test area in northern Russia may perhaps have given some hints on the action of SSC-X-9. Articles by David E. Sanger and Andrew E. Kramer (https://www.nytimes.com/2019/08/12/world/europe/russia-nuclear-acc ident-putin.html), David Axe (2019) and Ankit Panda (2019) explore features of these prospective new weapons.

Another long-term concern is that, if nuclear weapons were to be used, dust clouds resulting from the nuclear explosions could block the Sun's rays and cause a mini-Ice Age on Earth. As indicated in Fig. 7.3, the numbers of nuclear weapons that exist mean that even a restricted military action with the standard nuclear bombs stored and produced by the USA and Russia would still have an "overkill" potential to destroy cities and abolish human civilisation and other forms of life. An increasing number of small states have developed or begun to develop their own nuclear weapons. It is possible that terroristic organizations may do the same.

7.1.4 *Efforts to Restrict Nuclear Weapons Through International Treaties*

The Soviet Union developed its own nuclear war program very fast after the end of World War II. The largest nuclear fission bomb ever built was the Russian Tsar bomb, which was detonated in a test over Novaya Zemlya in October 1961. This bomb released the equivalent energy of 50 megatons TNT, 500-fold more than the energy of the Hiroshima bomb. Between 1952 and 1964, the UK, France and China also developed nuclear weapons, leading to more international tension about the possibility of the weapons being used in war. The testing of nuclear weapons above ground by several countries generated much concern about the effects of radioactive fallout in the atmosphere, especially after radioactive materials were identified in milk and grains in the USA. These concerns led, in 1963, to the first Nuclear Test Ban Treaty, initially signed by the USA, the Soviet Union and the UK and later by many other states. This Treaty banned further testing of nuclear weapons underwater, in the air, or in outer space, but allowed underground detonations to continue (https://www.jfk library.org/learn/about-jfk/jfk-in-history/nuclear-test-ban-treaty). A comprehensive nuclear test ban treaty was adopted at the United Nations in 1996, with the goal to ban all planned nuclear explosions in all environments. Unfortunately, this Treaty has yet to be ratified by all states (https://www.un.org/disarmament/wmd/nuclear/ ctbt/). Since 2012, the UN General Assembly has adopted a resolution to "take forward multilateral nuclear disarmament negotiations". A conference was held in 2017 to set forward a Treaty on the prohibition of nuclear weapons (http://undocs. org/A/CONF.229/2017/8). In October 2020, a landmark event was reached when the fiftieth UN nation ratified this Treaty. This important Treaty, with its potential to move nations towards complete nuclear disarmament, came into legal force in January 2021.

The International Atomic Energy Agency (IAEA) was set up in 1957 as a consequence of a speech in 1953 by USA President Eisenhower on "Atoms for Peace" and is part of the United Nations. It has the goals to "accelerate and enlarge the contribution of atomic energy to peace, health and prosperity throughout the world. It shall ensure, so far as it is able, that assistance provided by it or at its request or under its supervision or control is not used in such a way as to further any military purpose". It takes part in verification in connection with nuclear disarmament and arms control in UN member states and also takes part in major medical emergencies. Its present medium-term strategy (2018–2023) relates to increasing the peaceful uses of nuclear energy for nuclear power, promoting nuclear technology and its development (for example, for medical purposes) and improving nuclear safety and security, while continuing to deter proliferation of nuclear weapons (https://www.iaea.org/sites/def ault/files/16/11/mts2018_2013.pdf).

The Nuclear Non-Proliferation Treaty, which was set up in 1968, is an international treaty that aims to limit the expansion of nuclear weapons and works for cooperation on disarmament and peaceful applications of nuclear energy. 191 states are signatories and in 1995 it was decided to continue this Treaty indefinitely, with review every five years (www.un.org/disarmament/wmd/nuclear/npt/). Although some Review conferences have reached constructive outcomes, the 2015 conference did not reach a consensus. A Review conference was due in 2020 but has been postponed, apparently to August 2021, due to the covid-19 pandemic. It is intended to hold this conference by April 2021 (https://www.un.org/en/conferences/ npt2020). A general issue for the resilience of the Non-Proliferation Treaty is that not all countries that possess nuclear weapons, or have programmes to develop them, are signatories to this treaty.

In view of the very heavy risk for the future of the biological world, which could be toppled to destruction if nuclear weapons are unleashed, it is very surprising that nuclear weapons have not been banned. In 1987, the Intermediate-Range Nuclear Forces (INF) treaty, signed between the USA and the Soviet Union, was celebrated as the end of the Cold War. This treaty provided for a rather limited reduction of the number of short-range missiles including removal of cruise missiles from western European countries, banning of missiles with ranges between 500 and 5,500 km and made agreements for on-site inspections. It was a sign of hope for further disarmaments. Unfortunately, since 2014, the USA has considered that Russia has violated the terms for long-range missiles, through new missile developments such as discussed in Sect. 7.1.3. In 2019, the USA withdrew from the INF treaty (www.armscontrol. org/factsheets/INFtreaty). A remaining treaty between the USA and Russia is New Start, which ran from 2011 until 2021, and for which there is agreement (in progress at time of writing) to extend by a further five years. This Treaty places agreed limits on the number of warheads, intercontinental missiles or launchers that either country can hold, with a deployment limit of 1,550 strategic warheads each. It also includes agreed methods for verification including site visits and exchange of data (https://www.state.gov/new-start/).

The danger for life on Earth from a nuclear war is also increased by the strategy for atomic warfare which has persisted since the time of the USA/Soviet Union Cold War. The principle has been named Mutually Assured Destruction (MAD). We may take the scenario of a war between Russia and the USA as an example and let the USA shoot first to hit a Russian nuclear missile station and other targets in Russia. In the 5 min that the missiles take to cross the Atlantic they will be detected by Russian tracer systems. These would initiate an immediate counter-attack with nuclear weapons targeted on big cities in the USA. If the USA was thought to be under nuclear attack, its President has the sole authority to authorise missile launches by providing nuclear codes to the Pentagon. This type of war strategy is very open to misinterpretations, computing errors and/or hacking of computer systems (Guardian Briefing 2018). Thus, another aspect of reducing the risk of nuclear warfare could be to increase the protocol steps and times for decision-making procedures before any missile launch.

Indeed, stories are told of a last-minute prevention of a false counter-attack by a Russian officer. It is not clear how true these stories are. There are certainly many verified reports of fatal computer errors in non-military situations, particularly under the action of hackers and terrorists. An illustrative example comes from Switzerland, where huge amounts of money were spent for the development of an e-voting system. This was tested in several Cantons and was also given to professional hackers for testing. The hackers were able to produce any voting result that they liked. Consequently the e-voting system was withdrawn and millions of Swiss francs were lost (Fenazzi 2019).

Clearly, the current arsenal of nuclear weapons poses huge dangers. If released, nuclear weapons could, in the worst case, eliminate human life and ruin the entire biological world. The British astronomer Sir Martin Rees believes that the probability for humans to survive to the end of the present century is about 50% (Rees 2003). His extremely negative conclusion is based on a summation of all human-made risks. Ehrlich et al. (1983) expressed a similar view based on the risks from nuclear weapons alone. The Bulletin of the Atomic Scientists is an organisation founded by scientists in 1945 to provide information about man-made risks to human life. On their Doomsday Clock, midnight represents an ultimate catastrophe and the distance of the minute hand of the clock before midnight represents the level of present risk. In January 2020, the minute hand was re-set to 100s before midnight, and remains at this time for 2021, in view of the combined risks of the retreat from nuclear arms control and on-going global climate change. This is the closest the Doomsday Clock time has ever been to midnight; closer by 20s than during the Cold War (https://thebulletin.org/doomsday-clock/current-time/). Although in those days there were far more nuclear weapons in existence than at present, the modern weapons have greater capacities, as discussed in Sect. 7.1.3.

Other authors doubt these negative predictions, and have pointed out that, to date, only two small nuclear bombs have actually been used in a conflict and that nuclear weapons are kept under strict control. An example is Eric Gujer (2019), editor of a leading Swiss newspaper, who published an editorial "Logics of Scare". He mentioned two ways to solve the problem of nuclear warfare: either to strengthen the western European nuclear weapon shield under American guidance, or to seek a worldwide ban of nuclear weapons. He recommended the first solution. Several European politicians have discussed a universal ban, but he considers these moves to be unrealistic. Concerning the possibility of an effective global treaty to ban all nuclear weapons, he is unfortunately correct that this is probably impossible to achieve. We are convinced, however, that the simultaneous development of increasing numbers of ever more powerful and dangerous nuclear weapons is ultimately a way to a global disaster.

Many leading politicians point to the positive effect of nuclear deterrence to prevent conventional wars. It can be considered that under the influence of a broad political and military propaganda, society was appeased by the fact that no nuclear war has broken out. One close exception was the Cuban missile crisis in 1962, in which a nuclear war was almost initiated between the USA and the Soviet Union. When the INF disarmament contract was signed in 1987, hope for further reductions of nuclear weapons began to rise and the deterrence strategy appeared promising. It is hoped that the New Start treaty will be extended.

However, as outlined above, the INF contract has broken. An arms race like the one during the cold war has been initiated with different types of missiles. The USA has announced the development of medium size bombs (several times larger than the Hiroshima bomb) and Russia threatens with deadly new developments such as the Skyfall. In addition, North Korea (that withdrew from the Nuclear Non-proliferation Treaty) has a nuclear weapons programme and has been testing nuclear weapons since 2006. Many present national leaders who have the capacity to initiate nuclear wars take adversarial nationalistic positions. There are reports by experts that simple fission bombs could be built by terrorists, and that it is a question of time for this to happen. A risk estimate in 2013 calculated the 90% confidence interval of the annual probability of an accidental nuclear war breaking out between the USA and Russia to be between 0.001% and 7% (Barrett et al. 2013). Overall, with the weakening of historic treaties, efficient control of nuclear weapons seems an illusion. We can only repeat that nuclear wars carry the risk of abolishing humans and possibly life on Earth. This is a huge risk and may be called "human suicide and global murder". The only way to securely prevent this global disaster is to ban the production and storage of all nuclear weapons.

It is hard to see how this can be achieved practically under the present political and social systems. Many people are fully engaged with the daily needs of life and survival and can give little attention to the dangers of nuclear weapons. Others believe that everything is safe under the present treaties in spite of alarming evidence to the contrary. It is part of the human character that we tend to believe only those things that we have experienced directly. Fortunately, the world has not experienced nuclear warfare for more than 70 years. We believe that humans are intelligent enough to take serious counter-actions when they realize that a global disaster is next door.

7.2   Chemical Weapons

7.2.1   *Fritz Haber and His Followers*

The cruel philosophy of chemical agents in warfare is not to kill the enemy outright but to weaken them by pain, shock and disability and the need to care for the victims. To achieve this, many different poisonous agents have been synthesised chemically, and been used by military forces (Table 7.1).

The first military chemical attack was initiated by Germany during the First World War, using gases. The German chemist Fritz Haber planned and guided attacks against the French army that involved opening cylinders of chlorine gas that would in part be converted to hydrochloric acid near the French trenches. Another gas used was phosgene, which was supposed to penetrate the trenches and burn the lungs of the soldiers. Because of the primitive means of distributing the gas the military success (in other words the painful death of soldiers) was very dependent on the wind direction (Fig. 7.4).

Fritz Haber (Fig. 7.5) was an ingenious chemist who became a very controversial figure. He was born the son of Jewish parents in Breslau, on December 9, 1868. He studied chemistry in Heidelberg and Karlsruhe. He became very nationalistic and offered his services as a chemist to the German government. He also converted to the Christian religion. In 1901, Fritz Haber married Dr. Clara Immerwahr, the first woman to receive a Ph.D. degree in Chemistry at the University of Breslau in Germany (Fig. 7.5).

Along with many other discoveries in chemistry, Haber developed the Haber-Bosch method for the synthesis of ammonia from nitrogen gas and hydrogen at high pressure and by the action of a catalyst:

$$3H_2O + N_2O \rightarrow 2NH_3$$

This method was a fundamental breakthrough for the synthesis of ammonia-based fertilizers and in 1918 Haber received the Nobel Prize in Chemistry. Before this invention, fertilizers had to be produced slowly by decomposition of plant materials and animal products. The new method was first applied in German agriculture before the start of World War I and rendered the country independent of imports, which became very important when imports were later hindered by the war. Even today, most synthetic fertilizers are produced by the Haber-Bosch method. On the other hand, Fritz Haber introduced the first poisonous gases used in chemical warfare and became personally engaged in directing the gas war.

Table 7.1 Summary of chemical weapons and their actions (from Wikipedia, English version)

Classes of chemical weapon agents

Class of agent	Names	Mode of action	Signs and symptoms	Rate of action	Persistency
Nerve	• Cyclosarin (GF) • Sarin (GB) • Soman (GD) • Tabun (GA) • VX • VR • Some insecticides • Novichok agents	Inactivates enzyme acetylcholinesterase, preventing the breakdown of the neurotransmitter acetylcholine in the victim's synapses and causing both muscarinic and nicotinic effects	• Miosis (pinpoint pupils) • Blurred/dim vision • Headache • Nausea, vomiting, diarrhea • Copious secretions/sweating • Muscle twitching/fasciculations • Dyspnea • Seizures • Loss of consciousness	• Vapors: seconds to minutes; • Skin: 2–18 h	VX is persistent and a contact hazard; other agents are non-persistent and present mostly inhalation hazards
Asphyxiant/Blood	• Most Arsines • Cyanogen chloride • Hydrogen cyanide	• **Arsine:** Causes intravascular hemolysis that may lead to renal failure • **Cyanogen chloride/hydrogen cyanide:** Cyanide directly prevents cells from using oxygen. The cells then use anaerobic respiration, creating excess lactic acid and metabolic acidosis	• Possible cherry-red skin • Possible cyanosis • Confusion • Nausea • Patients may gasp for air • Seizures prior to death • Metabolic acidosis	Immediate onset	Non-persistent and an inhalation hazard

(continued)

Table 7.1 (continued)

Classes of chemical weapon agents

Class of agent	Names	Mode of action	Signs and symptoms	Rate of action	Persistency
Vesicant/Blister	• Sulfur mustard (HD, H) • Nitrogen mustard (HN-1, HN-2, HN-3) • Lewisite (L) • Phosgene oxime (CX)	Agents are acid-forming compounds that damages skin and respiratory system, resulting burns and respiratory problems	• Severe skin, eye and mucosal pain and irritation • Skin erythema with large fluid blisters that heal slowly and may become infected • Tearing, conjunctivitis, corneal damage • Mild respiratory distress to marked airway damage	• **Mustards**: Vapors: 4–6 h, eyes and lungs affected more rapidly; Skin: 2–48 h • **Lewisite**: Immediate	Persistent and a contact hazard
Choking/Pulmonary	• Chlorine • Hydrogen chloride • Nitrogen oxides • Phosgene	Similar mechanism to *blister agents* in that the compounds are acids or acid-forming, but action is more pronounced in respiratory system, flooding it and resulting in suffocation; survivors often suffer chronic breathing problems	• Airway irritation • Eye and skin irritation • Dyspnea, cough • Sore throat • Chest tightness • Wheezing • Bronchospasm	Immediate to 3 h	Non-persistent and an inhalation hazard

(continued)

Table 7.1 (continued)

Classes of chemical weapon agents

Class of agent	Names	Mode of action	Signs and symptoms	Rate of action	Persistency
Lachrymatory agent	• Tear gas • Pepper spray	Causes severe stinging of the eyes and temporary blindness	Powerful eye irritation	Immediate	Non-persistent and an inhalation hazard
Incapacitating	• Agent 15(BZ)	Causes atropine-like inhibition of acetylcholinein subject. Causes peripheral nervous system effects that are the opposite of those seen in nerve agent poisoning	• May appear as mass drug intoxication with erratic behaviors, shared realistic and distinct hallucinations, disrobing and confusion • Hyperthermia • Ataxia (lack of coordination) • Mydriasis (dilated pupils) • Dry mouth and skin	• Inhaled: 30 min–20 h; • Skin: Up to 36 h after skin exposure to BZ. Duration is typically 72–96 h	Extremely persistent in soil and water and on most surfaces; contact hazard
Cytotoxic proteins	Non-living biological proteins, such as: • Ricin • Abrin	Inhibit protein synthesis	• Latent period of 4–8 h, followed by flu-like signs and symptoms • Progress within 18–24 h to: • Inhalation: nausea, cough, dyspnea, pulmonary edema • Ingestion: Gastrointestinal hemorrhage with emesis and bloodydiarrhea; eventual liver and kidney failure	4–24 h; see *symptoms*. Exposure by inhalation or injection causes more pronounced signs and symptoms than exposure by ingestion	Slight; agents degrade quickly in environment

Fig. 7.4 Release of chlorine gas into the French army trenches by the German army during the first World War (from Wikipedia, English version)

Fig. 7.5 Fritz Haber, 1918. Clara Immerwahr, the first wife of Fritz Haber (both from Wikipedia, English version)

7.2.2 First Opposition Against Chemical Warfare by Haber's Wife

Clara Immerwahr is thought to have condemned the use of poison gases in war as a "perversion of the ideals of science" (Von Leitner 1994; Friedrich and Hoffmann 2017). Shortly after the first chlorine gas attack took place at Ypres on 22 April, 1915, and on the evening of a victory celebration, she shot herself with the army pistol of her husband. Very many people including the authors of this book find her life-story very touching.

On several occasions Fritz Haber justified his engagement with gas attacks by his wish to shorten the war through a German victory. In 1933, because of his Jewish past, Fritz Haber had to retire from the Max Plank Institute that had been named after him. He received an appointment at the University of Cambridge, England. Shortly afterwards, he accepted an offer by Chaim Weizmann to direct the Weizmann Institute of Science in Rehovot, Israel. On his journey to Israel he died in a hotel in Basel at the age of 65 years. He is buried at the Hörnli churchyard in Basel. At the request

Fig. 7.6 Stamp of the German Post from the 1957 series, "Men from the history of Berlin"

of his son, the urn of his first wife Clara was transferred to the same burial place. It is much appreciated that the neighbouring German city of Basel Lörrach named a street after Clara Immerwahr to honour her denunciation of gas attacks.

Judgements on Fritz Haber are manifold and very controversial. His role as the father of gas warfare must be viewed along with his other achievements in chemistry and the increases to agricultural productivity. The Fritz Haber Max Plank Institute in Berlin still continues and has an excellent reputation. The German post office has issued a postage stamp in his memory (Fig. 7.6).

Several authors have been stimulated to write books or direct television features and movies about Fritz Haber and his life: for example the books by Szöllözi-Janze (1998), Wille (1996), Thiessen (2003). The most objective evaluations of the relationship of Fritz Haber and Clara Immerwahr can be found in a recent podcast by Gudrun Kammasch (2014) and an article by Friedrich and Hoffmann (2017) in a book "One Hundred Years of Chemical Warfare: Research, Deployment, Consequences". The BBC has broadcast several illuminating features in 2001 and 2008, and a movie "Haber" by Ragussis (2008), has won several prizes.

7.2.3 Uses of Chemical Weapons

Chemical weapons offered the first method for mass destruction. A military advantage of chemical warfare is the relative cheapness and ease in producing the weapons. However, the effects do not reach the killing potential of nuclear weapons, as radioactivity is not involved. In conventional military actions, the practical difficulties of targeting and directing chemical weapons only towards enemy troops has limited their use. These are brutal weapons. Whoever has seen the victims of a poison gas attack will suffer from what they has experienced for all their life. During the second World War, chemical weapons were not used in military actions, yet nerve gases were used by the Nazis for killing Jews and other groups in concentration camps.

Chemical weapons were used extensively in the war between Japan and China, in the Iran–Iraq War (1983–1988), and, most recently, during the years of war in Syria. The devasting effects of chemical weapons are illustrated by information from the Iran-Iraq war and the reported scale of casualties, as shown in Table 7.2. In addition to the effects on humans, these poisons will also have destroyed or maimed other living organisms.

7.2.4   Own Experience

Several years ago one of the authors (J.E.) had to serve in the Civil service of Switzerland. As a chemist and biologist he was put into the unit for protection against attacks by poisonous gases. Near the lake of Interlaken, a civil service laboratory had been established, Labor Spiez (Fig. 7.7), with the task of protecting the Swiss population against potential attacks by chemical weapons. The laboratory developed a high reputation in analysing the broad repertoire of chemicals with which human kill each other or are planning to poison their enemies. The Spiez Lab was active in investigating the gas attacks by Saddam Hussein against the Kurdish minority in Iraq. More recently, it appeared in the headlines when the Dutch government expelled two alleged Russian spies. They were accused of planning to hack into the Spiez laboratory, where Novichok nerve agent samples from the chemical attack in Salisbury in the UK had been analysed. In this attack, Sergei Skripal, a retired Russian military officer and his daughter, Yulia, who was visiting him at his home in Salisbury, were poisoned on 4 March 2018. The Spiez laboratory subsequently confirmed the British claim that the Skripals had been exposed to the military-grade nerve agent Novichok.

From the Spiez laboratory, the unit in which Jürgen Engel served obtained movies of attacks with mustard gas and the nerve gas tabun from the war by Saddam Hussein against the Kurds. Watching these movies was the most horrible experience in the life of Jürgen Engel. This was not only killing of people and children but planned torture with unbelievable pains. Jürgen Engel watched the movies with naked children full of blisters helplessly crying from terrible pain about 30 years ago. Writing this text, the scenes return to him now and he woke up several times at night from terrible dreams. He remembers how he would leave the civil service lectures in the evening. Walking home through the lively streets of Basel he could not believe that this kind of warfare is part of our human life.

Table 7.2 Iraqi chemical attacks during the Iran–Iraq War (from Wikipedia, English Version, 2020)

Date	Event	Location	Type	Casualties[a]
1983, August		Haj Umran	mustard	less than 100 Iranian/Kurdish casualties
1983, October–November		Panjwin	mustard	3,000 Iranian/Kurdish casualties
1984, February–March		Majnoon Island	mustard	2,500 Iranian casualties
1984, March	Operation Badr	al-Basrah	tabun	50–100 Iranian casualties
1985, March	Battle of the Marshes	Hawizah Marsh	mustard and tabun	3,000 Iranian casualties
1986, February	Operation Dawn 8	al-Faw	mustard and tabun	8,000–10,000 Iranian casualties
1986, December		Um ar-Rasas	mustard	1,000 Iranian casualties
1987, April	Siege of Basra (Karbala-5)	al-Basrah	mustard and tabun	5,000 Iranian casualties
1987, June	Chemical bombing of Sardasht	Sardasht	mustard	8,000 Iranian civilians exposed
1987, October		Sumar/Mehran	mustard and nerve agent	3,000 Iranian casualties
1988, March	Halabja chemical attack	Halabjah and villages around Marivan	mustard and nerve agent	1,000 Iranian Kurdish civilian casualties
1988, April	Second Battle of al-Faw	al-Faw	mustard and nerve agent	1,000 Iranian casualties
1988, May		Fish Lake	mustard and nerve agent	100 or 1,000 Iranian casualties
1988, June		Majnoon Island	mustard and nerve agent	100 or 1,000 Iranian casualties
1988, May–June		villages around Sarpol-e Zahab, Gilan-e-gharb and Oshnavieh		Iranian civilians

(continued)

Table 7.2 (continued)

Date	Event	Location	Type	Casualties[a]
1988, July		South-central border	mustard and nerve agent	100 or 1,000 Iranian casualties

[a]The actual casualties may be much higher, as the latency period is as long as 40 years (Hersey 1946b)

Fig. 7.7 The Swiss centre for the analysis of poisons used in chemical warfare

7.2.5   *Organisation for the Prohibition of Chemical Weapons*

During the history of the United Nations several attempts have been made to put chemical warfare onto the register of forbidden war activities. The first attempts were as unsuccessful as other restrictions of war activities. Although illegal, killing of civilians, bombing of civilian areas in cities and even destruction of hospitals and schools continues in warfare. We have to be very grateful to the Netherlands for taking a, to date, successful initiative in 1997 in response to the international shock caused by gas attacks in Syria. Under the guidance of the Netherlands, the Organisation for the Prohibition of Chemical Weapons (OPCW) that currently includes 193 member states was established with headquarters in The Hague (www.opcw.org). The OPCW has an overall mission to eliminate chemical weapons and runs effective control systems including tests for the precursor chemicals used to make chemical weapons. It includes anti-spying and anti-cyberattack offices. These efforts prevented two attacks involving the Russian Main Intelligence Directorate (GRU). According to news on Swiss television in October 2018, four officers of GRU were deported from the Netherlands for hacking into the office of OPCW. Two officers of the same organization were deported for an attempt to break into the information system of the Swiss Spiez Laboratory (see Sect. 7.2.4). Although it does not have judicial powers, OPCW seeks to identify those responsible for chemical attacks. In April 2020, the new Investigation and Identification team of OPCW released its first report after investigating attacks that used the chemical weapons sarin and chlorine in Ltamenah,

Syria, in March 2017 (https://www.opcw.org/sites/default/files/documents/2020/04/s-1867-2020).

We must hope that the OPCW will continue to restrict future activities in chemical warfare. In 2013, the organization was awarded the Nobel Peace Prize. In the laudation, the Norwegian Nobel Committee chairman Jagland said: "The conventions and the work of the OPCW have defined the use of chemical weapons as a taboo under international law". Chemical weapons have horrible effects and, to date, the largest attacks have been reported to have caused more than a million causalities. Far greater numbers of casualties are possible in nuclear war, along with widespread, long-lasting and severe effects on other living organisms. Hopefully, therefore, similar measures can be taken against nuclear warfare as well.

The use of biological weapons is also forbidden under international treaties (the last being the Biological Weapons Convention of 1972), yet concerns of modified infectious agents being used in a bioterrorism attack or by a rogue state are increasing. As with chemical and nuclear weapons, the effects of such agents, if used, cannot be contained easily. There is a long history of use of infectious materials in warfare (Frischknecht 2003). The efforts to control rabbit numbers in Australia by using a virus (see Chap. 6) can be taken as an example of the types of methods humans could possibly use to conduct biological warfare either directed at humans, or to cause starvation by destroying animals or plants.

References

Axe D (2019) Why Russia's nuclear-powered "skyfall" missile is bad news, bad for the environmental and global security. The National Interest, October 22. https://nationalinterest.org/blog/buzz/why-russias-nuclear-powered-skyfall-missile-bad-news-90116

Barrett AM, Baum SD, Hostetler K (2013) Analyzing and reducing the risks of inadvertent nuclear war between the United States and Russia. Sci Glob Secur 13:106–133. https://doi.org/10.1080/08929882.2013.798984

Ehrlich PR et al (1983) Long term biological consequences of nuclear war. Science 222:1293–1300

Fenazzi S (2019) E-Voting auf Eis: Was bleibt von den Bemühungen? Swissinfo.ch, Juni 27. https://www.swissinfo.ch/ger/politik/schweiz-demokratie-volksabstimmungen-evoting/45061040

Friedrich B, Hoffmann D (2017) Clara Immerwahr: a life in the shadow of Fritz Haber. In: Friedrich B, Hoffmann D, Renn J, Schmaltz F, Wolf M (eds) One hundred years of chemical warfare. Springer, Cham

Frischknecht F (2003) The history of biological warfare. EMBO Rep. 4(Suppl 1):S47–S52. https://doi.org/10.1038/sj.embor.embor849

Grujer E (2019) Experten für den Weltuntergang (editorial) Neue Zürcher Zeitung 240, 97, April 27

Guardian Briefing (2018) https://www.theguardian.com/world/2018/jul/16/nuclear-war-north-korea-russia-what-will-happen-how-likely-explained

Gujer E (2019) Die Logik des Schreckens Neue Zürcher Zeitung 240 (33) February 9

Hersey J (1946a) A reporter at large. Hiroshima. New Yorker Magazine, August 31, 1946

Hersey J (1946b) Hiroshima. Penguin Books

Kammasch G (2014) Fritz Haber-Clara Immerwahr, Was lehrt uns die Geschichte. YouTube, UiniHeidelberg, Studium Generale

Panda A (2019) The absurd strategy behind Russia's nuclear explosion. The New Republic, August 2019. https://newrepublic.com//article/154815/absurd-strategy-behind-russias-nuclear-explosion

Paul J (1837) Sämmtliche Werke, Band 4, Der kleine Krieg in der Brust and Dämmerungen, vol 4. Tetot Freres, Paris, pp 1–17

Pontin J (2007) Oppenheimer's ghost. Technology Review November to December

Ragussis, D (2008) Haber. A short film, see https://www.imdb.com/title/tt1258199/

Rees M (2003) Our final century, will civilisation survive the twenty-first century? Willian Heinemann Ltd.

Szöllösi-Janze M (1998) Fritz Haber 1968–1934. Beck, München

Thiessen V (2003) Einstein's gift. Canada Press, Playwrights

Von Leitner G (1994) Der Fall Clara Immerwahr: Leben für eine humane Wissenschaft. Beck, München

Wille HH (1996) Der Januskopf Velag Neues Leben

Chapter 8
Human-Made Risks and Climate Change with Global Heating

Throughout their history, humans have modified their environment. Remains of Neolithic buildings, and monuments such as the Pyramids or Stonehenge are thousands of years old. Even by 3,000 years ago, human hunting and agricultural activities had wrought changes on natural landscapes (AchaeoGLOBE Project 2019). Over the last few hundred years humans have had increasingly huge effects on the natural world. These changes have taken place in a very short time compared with the time periods over which geological changes and biological evolution act. The high rate of recent change is demonstrated by experiences within our lifetimes. Villages of our childhoods have grown to suburbs of large cities; rail, road and air traffic have expanded; wildlife areas have disappeared; open lands and forests have been replaced by power stations, industrial developments, and airports. Globally, human-made objects now out-weigh the total mass of all living organisms on Earth (Elhacham et al. 2020).

Humans have also caused the extinction of many species through excessive hunting or slaughter for commercial purposes: the giant auk and the Tasmanian tiger (thylacine) are evocative examples. Over the twentieth century, increasingly widespread use of pesticides further contributed to reductions in populations and loss of species, especially amongst insects. While purposeful conservation efforts to re-introduce charismatic species after local extinctions have been successful (for example, the European bison in central Europe or the osprey in England) these projects are on a small scale compared to the current rapid decline of many species. It has also become clear that human activities are changing the climate of the whole planet, with potentially dire consequences for human civilisation and biodiversity in general. In this chapter, we discuss the evidence for climate change through global warming and the evidence that this global heating has an anthropogenic origin. We discuss the inter-linked issue of the increasing human population and then turn to the ongoing efforts to mitigate or arrest climate change. Finally, we will discuss why massive efforts to decrease greenhouse gas emissions are needed so urgently. This area has also become a major political issue.

J. C. Adams and J. Engel, *Life and Its Future*,
https://doi.org/10.1007/978-3-030-59075-8_8

8.1   Evidence for Global Warming

To make any statement about whether the world climate is, in general, becoming warmer, the first step is to have data on historic temperatures around the world. Estimates of land or ocean temperatures in historic times are made based on so-called temperature proxies. For example, there are chemical differences in snow formed at different temperatures, so core samples from the polar ice sheets can provide estimates of temperatures going back over thousands of years. The chemistry of fossil shells in marine sediments (from organisms such as diatoms) provides information on ocean temperatures over time. According to such paleoclimatology data, in the long-ago Pliocene era (~2–5 million years ago) the Earth's climate was hotter than in the present Holocene era (Fig. 8.1; the Holocene corresponds to the last 10,000 years, beginning after the last Ice Age). Throughout the last 800,000 years there have been only two periods that were warmer than the Holocene (Fig. 8.1, lower panel). The relative warming at these times was less than + 1 C, yet in these periods sea levels were much higher than at present (Hansen and Sato 2012). The Holocene era appears to have started off relatively warm with maximum temperatures (to date) being experienced between 9000 and 5000 years ago, followed by intermittent warming and cooling trends and a Little Ice Age between 1550 and 1850 (Renssen et al. 2009; Shakun 2018). However, new data indicate that the world at present is at the hottest it has been for the last 12,000 years (Bova et al. 2021).

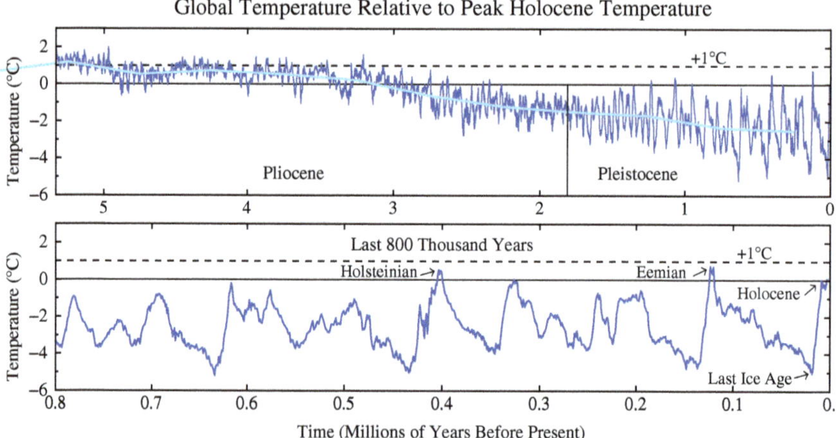

Fig. 8.1 Global average temperature plotted relative to peak temperature in the Holocene, based on core samples from ocean sediments. The upper panel covers the last 5 million years and the lower panel shows the last 0.8 million years only. Note that '0' time on the x axis does not accurately represent up to the present day. Fig 8.2 provides an accurate chart of the global average temperature anomaly over pre-industrial levels during the last 140 years. (From https://www.giss.nasa.gov/res earch/briefs/hansen_15/)

With regard to global temperatures throughout the last 150 years, we have the benefit of direct observations of weather conditions, including temperature recordings, which have been a major area of scientific endeavour. The UK Meteorological Office was set up in 1854 by Robert Fitzroy, the former Captain of the *HMS Beagle* (see Chap. 4). In 1873, the International Meteorological Organisation was founded and began to set common standards and methods for temperature recording by different countries. Thanks to this history of detailed weather recording from many weather stations and ships, a huge number of measurements of air temperature over land and water temperature at the ocean surface are available. From these, the global average temperature over the last ~100 years can be calculated. In these calculations, averaging over 30 year periods is used to take account of natural year-to-year variations that may occur (ipcc.ch).

These records show a gradual upward trend in the global average temperature from about 1920 onwards and clear evidence that a sharp increase in the global average temperature began in the 1960s. In Fig. 8.2, this temperature change is displayed as the temperature anomaly relative to the pre-industrial average temperature. It is important to note that, in the context of these calculations, "pre-industrial" refers to the period 1850–1900, for which climate recordings are available from many parts of the world, not to the true pre-industrial period. In 2020, the temperature anomaly stood at +1.16 °C (Global Climate Report—February 2020). To put this in terms of the actual temperatures, around 1900 the global average temperature was ~13.7 °C and in 2019 it was 14.85 °C (USA National Oceanic and Atmospheric Administration, State of the Climate Report 2020; https://www.ncdc.noaa.gov/sotc/global/202002).

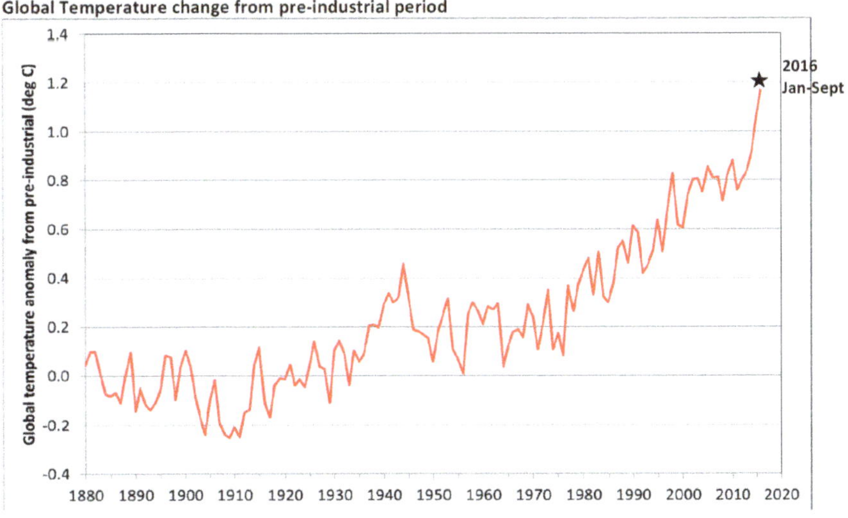

Fig. 8.2 The global average temperature anomaly, compared to the pre-industrial era, for the period 1880 to 2016. From the World Meteorological Organisation Statement on the Status of the Global Climate in 2016, based on data from NOAA, NASA, UK Met Office/Climatic Research Unit

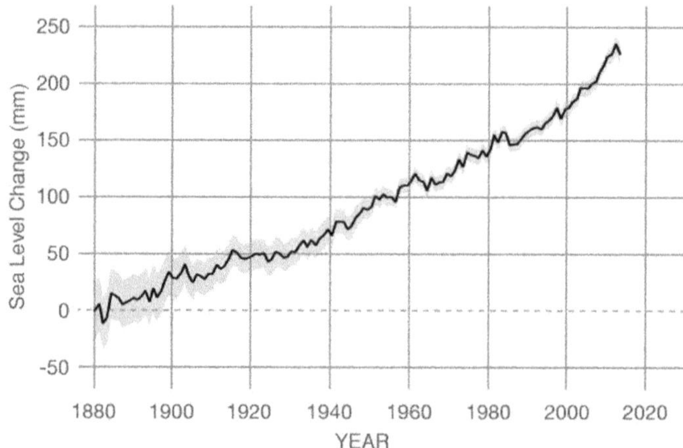

Fig. 8.3 The increase in sea level during the period 1880–2013. Data from coastal tide gauge records (*Credit* Information and Data Centre, Commonwealth Scientific and Industrial Research Organisation)

The phenomenon of increased global average temperature is referred to as global warming. The hottest month on record to date was July 2019, for which the global average temperature was 0.95 °C above the twentieth century average (https://www. noaa.gov/news/july-2019-was-hottest-month-on-record-for-planet). The year 2020 was the joint (with 2016) hottest year ever recorded (Copernicus Climate Change Service 2021).

The oceans comprise 71% of the Earth's surface and have absorbed much of this increase in the surface temperature. As reported in 2017, the temperature of the top 700 m of the oceans has increased by an average of +0.2 °C since 1969 (Levitus et al. 2017). In parallel, tide gauge records show that sea levels have been rising steadily over the last century (Fig. 8.3). During the last twenty years the rate of increase has doubled (Nerem et al. 2018), such that sea levels are currently rising at an average of 3.3 mm per year (https://climate.nasa.gov). The rise in sea level is partly accounted for by the thermal expansion of water due to the increase in ocean temperature and partly by water that is being added to the oceans due to melting of the polar ice caps, sea ice and ice on mountains and glaciers.

The shrinkage of polar ice sheets, both in the Antarctic and in Greenland, is very clear and has been documented since 2002. It is calculated that these losses have added 14 mm to the overall sea level throughout this period (Smith et al. 2020). Melting of the ice sheets of the Antarctic is calculated to have contributed 8 mm to sea level since 1992 (Bell and Seroussi 2020). Each year, the Greenland ice sheet is losing 280 Gigatonnes of ice (Fig. 8.4), which is calculated to contribute 0.7 mm of the global increase in sea level each year. In 2019, a record 532 Gigatonnes of ice was lost (Sasgen et al. 2020; Velicogna et al. 2020) and in, 2020, 43% of the Milne Ice Shelf in the Canadian Arctic was lost through fracture of the Shelf (https://www.nes dis.noaa.gov/content/canada%E2%80%99s-milne-ice-shelf-collapses). For context, if the entire Greenland ice-sheet melted, sea levels would rise by 7 m, drowning large

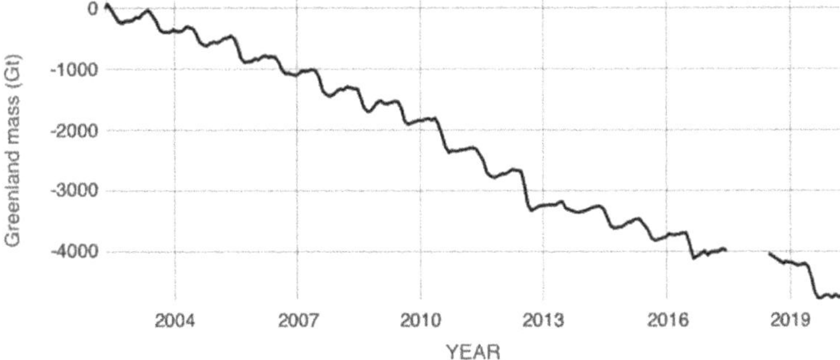

Fig. 8.4 The loss of Greenland's ice sheets from 2002 onwards. Measurements from NASA's GRACE and GRACE Follow-On satellite images of the land ice sheets in Greenland since 2002. The gap for 2017 is because GRACE Follow-On began data collection in 2018 (https://www.nasa.gov/topics/earth/features/pia16294.html) (*Source* climate.nasa.gov)

areas of the world's land masses as we know them. Glaciers around the world are also shrinking, especially in the Himalayas (Xu et al. 2009) and the European Alps (Milner et al. 2017). The high elevations of these mountain regions mean that these areas are heating up faster than the global average. In addition, because of increased air pollution from human activities, black carbon particles (soot) are being deposited onto the ice and snow and the darkening of the surfaces also accelerates the rates of thawing (IPCC 2019). Overall, the melt-water contributes more water to rivers and, ultimately, the oceans (Tranter and Brown 2017).

These major changes in the physical environment have follow-on consequences that are also part of global warming. As ice-sheets melt, the darker surfaces of water or land that are revealed absorb more heat and act as positive feedbacks that amplify and drive further increases to global temperature: an effect known as the albedo feedback (Hansen and Sato 2012). With the loss of ice and snow, the Arctic and boreal permafrost below the land surface is beginning to warm up: there is some evidence that more carbon dioxide and methane are being released, which adds to the greenhouse gases (discussed in Sect. 8.2) in the atmosphere (IPCC 2019). Global mean atmospheric methane levels have been rising since 2007, and this is thought to result from a combination of processes as well as permafrost melt (Mikaloff-Fletcher and Schaefer 2019). Because of its high latitude, the Arctic region is warming faster than the global average, which means that increasingly high temperatures are being reached in the summer, which drive wildfires and contribute to the rate of permafrost melting. An increased influx of cool water from the Arctic region (due to the melting ice) into the oceans has been modelled to affect the strength or course of the Gulf Stream that brings warm water from the southern USA East coast across the North Atlantic to the UK and Scandinavia. The consequences of such a change remain under debate—many studies have predicted that Western Europe would become cooler, yet a recent study predicts that slowing of the Gulf Stream would tend to boost air temperature over the ocean, and so contribute to overall global warming (Chen and Tung 2018). We discuss other consequences for life on Earth in Sect. 8.4.

8.2   What Are the Physical Causes of Global Warming?

We discussed in Chap. 3 how the atmosphere formed around the early Earth and has a role in protecting the Earth's surface from extremes of heat and radiation from the Sun. Later, the activities of photosynthetic organisms added oxygen to the Earth's atmosphere. The atmosphere of the Earth is now made up, by volume, of 78% nitrogen, 21% oxygen, 0.93% argon, 0.04% carbon dioxide and trace amounts of neon, helium, methane, krypton, hydrogen and nitrous oxide. Water vapour makes up between 0 and 4% of the atmosphere by volume, depending on the time of day and location. Half of the atmosphere is present in a layer approximately 10 miles deep adjacent to the Earth's surface and this layer contains most of the water vapour.

The French mathematician Joseph Fourier (1768–1830) was the first to model that the Sun is the main source of energy that warms the Earth's surface and leads to emission of heat by the Earth, which he termed "dark heat". We now know that "dark-heat" is infra-red radiation. Fourier also postulated that properties of the atmosphere around the Earth act to trap the heat that is emitted. The mechanism for this was later identified by the physical chemist, John Tyndall (1820–1893) (Fig. 8.5), a pioneer in studying the properties of gases to absorb and transmit radiant heat. He discovered that water vapour, carbon dioxide, and methane, even though minor components of the atmosphere, are the strongest absorbers of heat. Tyndall's experiments provided the first scientific evidence on how the atmosphere around the Earth reduces heat loss from the Earth to space (Hulme 2009). In the early twentieth century this heat-trapping property became termed the "greenhouse effect".

Fig. 8.5 John Tyndall (From Wikipedia, English version)

Although the carbon stored in the Earth's natural atmosphere is only a tiny fraction of the total carbon associated with the Earth, changes in atmospheric carbon have a disproportionately large effect on the carbon cycle (see Chap. 3 for introduction of the carbon cycle), because of the roles of carbon monoxide and carbon dioxide as greenhouse gases. Since the mid-twentieth century, the amount of carbon dioxide in the atmosphere has been measured continuously at sites around the world. The longest established monitoring site is the Mauna Loa Volcanic Observatory, located in the Pacific Ocean at an elevation of 3400 m, a site where industrial pollution might be expected to be limited. Nevertheless, all the recording sites show the same trend of a continuous increase in the amount of carbon dioxide in the atmosphere, from ~310 parts per million (ppm) in 1958 to ~416 ppm in 2020 (Fig. 8.6).

Carbon dioxide is the most abundant greenhouse gas (Fig. 8.7). The increase in atmospheric carbon dioxide provides clear evidence that the capacity of the atmosphere to retain the Earth's radiant heat is increasing.

The global warming effect of greenhouse gases also depends on how long a gas remains in the atmosphere. Scientists have therefore developed a metric, the Global Warming Potential (GWP), to allow comparisons between different greenhouse gases. Carbon dioxide is given a reference value of 1; in fact, carbon dioxide will continue in the atmosphere for thousands of years. Methane, which absorbs more heat than carbon dioxide, has a GWP of 28–36. Nitrous oxide, that is more stable than methane, has a GWP of 265–298 (https://www.epa.gov/ghgemissions/unders tanding-global-warming-potentials). Both of these gases are also increasing in the atmosphere (Fig. 8.8) and, given their chemical and physical stability, they act as drivers of global warming. Nitric oxide is also present in the atmosphere (as a result of human activities, see Sect. 8.3) and is rapidly converted to nitrogen dioxide; these oxides of nitrogen have a GWP similar to that of methane. Another gas to consider is water vapour. The air holds more water vapour as its temperature increases. Since water vapour has greenhouse properties, this means that as the global average temperature increases, the greenhouse activity of water vapour is increased along with the

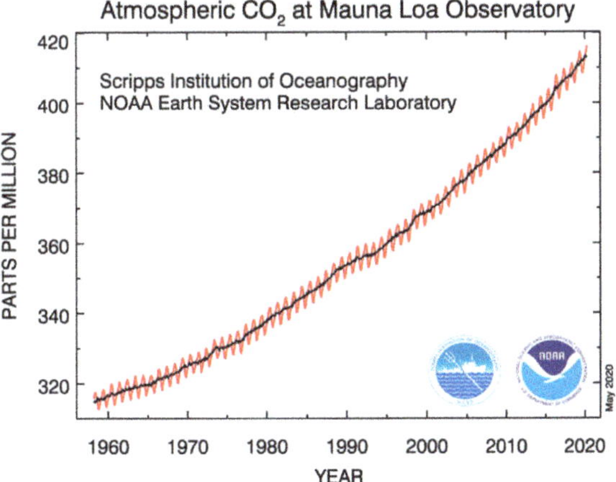

Fig. 8.6 The increase in atmospheric carbon dioxide levels between 1958–2020. Records of Mauna Loa Volcanic Observatory. The black line indicates the annual mean (*Credit* USA National Oceanic and Atmospheric Administration [NOAA] and Scripps Institution of Oceanography [https://research.noaa.gov/art icle/ArtMID/587/ArticleID/ 2636/Rise-of-carbon-dio xide-unabated])

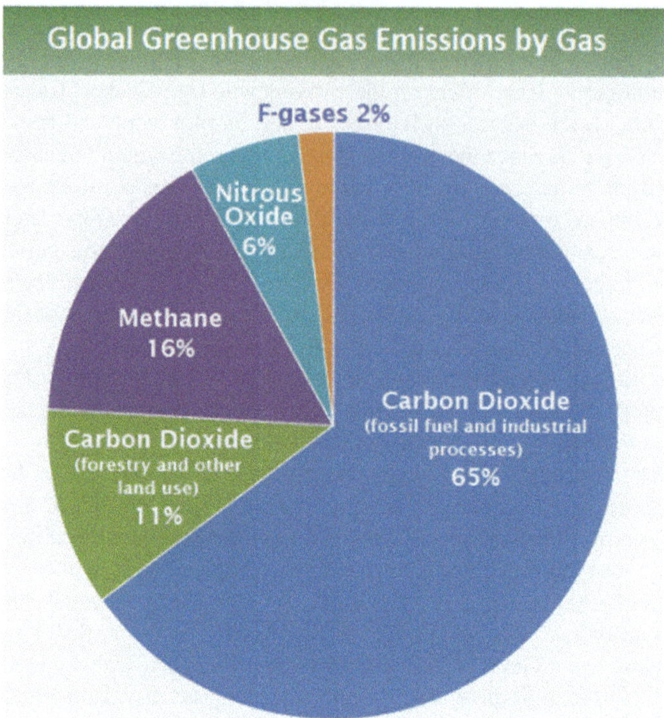

Fig. 8.7 Piechart of the major global greenhouse gas emissions. F-gases (fluorinated gases) cover synthetic gases such as hydrofluorocarbons. From USA Environmental Protection Agency, based on data in the IPCC report (2014) on emissions in 2010. In 2019, total global emissions were around 33 gigatonnes of CO_2 equivalents (https://www.epa.gov/ghgemissions/global-greenhouse-gas-emissions-data)

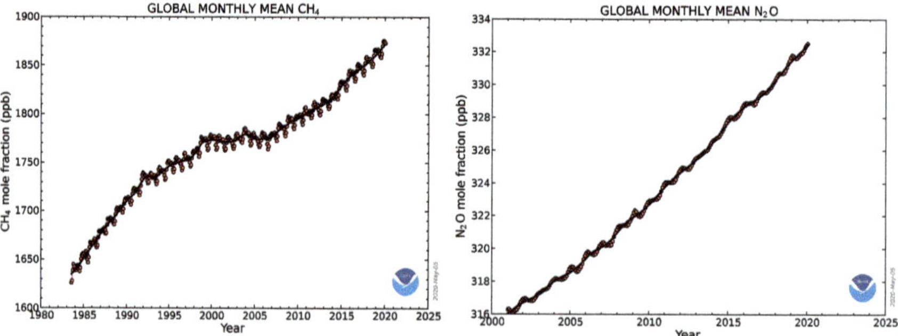

Fig. 8.8 The increase in atmospheric methane and nitrous oxide between 1984-2020 (methane) or 2002–2020 (nitrous oxide) (From Ed Dlugokencky, NOAA/ESRL [www.esrl.noaa.gov/gmd/ccgg/trends_ch4/ and www.esrl.noaa.gov/gmd/ccgg/trends_n2o/])

likelihood of it precipitating as rain. Water vapour thus has a positive feedback role in global warming and the likelihood of severe storms.

8.3   Evidence for an Anthropogenic Origin of Global Warming

Throughout much of Earth's history, natural factors alone have determined the composition of the Earth's atmosphere. As part of the development of human civilisation, humans learned to use fire in controlled ways for cooking, warmth and working in metals or ceramics; the evidence for human use of fire goes back 1.5 million years (Gowlett 2016). The burning of wood as a fuel began the process of human-mediated changes to the atmosphere, through the release of its stored carbon as carbon dioxide. Much later, between 1760 and 1830, the Industrial Revolution began in the UK, followed by Europe and North America. Until this time, manufacturing procedures were based on hand crafts that often used wind or water power, and most people lived in rural areas. The Industrial Revolution brought drastic technological and lifestyle changes in which people moved into cities to take part in manufacturing by large scale processes that depended on machines in factories.

Many of these machines were powered by steam engines, which ran on coal as a fuel to heat water to produce the steam. The first steam engines were invented by Thomas Newcomen in 1712 and were used to pump water out of coalmines. These engines were inefficient and needed vast amounts of coal to run. The widespread adoption of steam engines to power manufacturing processes took place as a result of improvements to their efficiency. A key innovation dates to 1769, when the Scottish inventor James Watt patented the design of a separate condensing chamber that reduced the loss of steam, thus greatly cutting fuel consumption. Later, Watt became a business partner with Matthew Bolton who owned the Soho Foundry in Smethwick, near Birmingham. With the condenser and additional innovations, the Bolton-Watt steam engine (Fig. 8.9) used about 75% less fuel than the Newcomen engine and became a mainstay for the mechanisation of industry (Allen 2017). The Industrial Revolution marked the beginning of modern times in which use of fossil fuels: coal, and more latterly oil and natural gas, became so extensive and essential to many human societies and the economies of many countries.

It is clear that the timing of the sharp increase in atmospheric carbon dioxide and other physical changes (Sect. 8.1) correlates with the time period over which humans greatly increased their technological activities that release greenhouse gases. Paleoclimatic measurements provide evidence for warming of tropical oceans that began in the nineteenth century (Abram et al. 2016). Nevertheless, how can we be sure that the strong increase in atmospheric carbon dioxide and global warming are due to human activities and not to variations caused by natural processes? For example, it has been considered that an increase in the energy output of the Sun might be driving global warming. However, if the heating was provided by the Sun, all the layers of

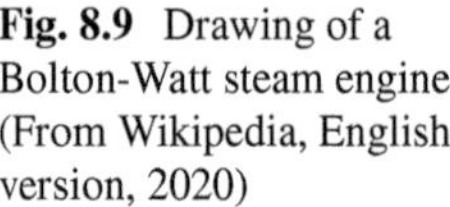

Fig. 8.9 Drawing of a Bolton-Watt steam engine (From Wikipedia, English version, 2020)

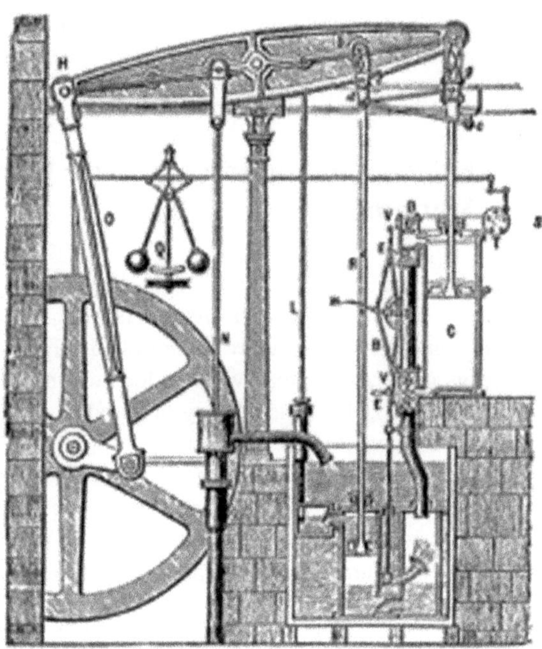

the atmosphere would be warmed, whereas measurements from satellite instruments show a cooling in the outermost atmosphere and warming only in the layer nearest the Earth. Also, the Sun's energy output has varied by no more than 0.1% over the last forty years, which is not sufficient to account for the rapid, continuing increase in global average temperature documented over this time period. These facts indicate that increased solar energy does not account for the current global warming (Fahey et al. 2017).

A separate line of evidence comes from the relative abundance of carbon isotopes in the atmosphere. Carbon exists as three isotopes that undergo the same chemical reactions but differ in mass: ^{12}C, ^{13}C and ^{14}C. Fossil fuels lack ^{14}C and, because they derive from plant sources (which take up ^{12}CO$_2$ in preference to ^{13}CO$_2$), contain less ^{13}C than the atmosphere. From measurements of the amount of each isotope in air samples, researchers have identified that the proportions of ^{13}C and ^{14}C relative to ^{12}C in the atmosphere have dropped over the same time period in which atmospheric carbon dioxide levels have increased. This is clear evidence that the burning of fossil fuels is causing the increase in atmospheric carbon dioxide (Keeling et al. 1979).

Other evidence that a natural origin of the current global warming is very unlikely comes from the unusual uniformity of the climate change processes that have been documented in different parts of the world over the last forty to fifty years. For more than 98% of the world, the warmest period within the last two thousand years occurred within the twentieth century. This global synchronicity is very different to what is known of historic, naturally occurring episodes of climate change, again indicating that a natural origin of the current global changes is very unlikely (Neukom et al. 2019a).

The fifth assessment report of the Intergovernmental Panel on Climate Changes (IPCC) in 2014 concluded that it is *extremely likely* that the influence of humans has been the dominant reason for global warming between 1900 and 2013 and that increased emissions of greenhouse gases are the most likely cause. This conclusion was re-stated in the 2018 IPCC report and is agreed upon by almost all climate scientists (Cook et al. 2016) and scientific professional bodies, for example the Royal Society of the UK and the USA National Academy of Sciences (National Academy of Sciences 2020). Deeper analyses of the historical climate, extending back over the last two thousand years, have concluded that it is *99% likely* that the current global warming and climate change is being caused by human activities (Brönnimann et al. 2019; Neukom et al. 2019a, b).

Paleoclimatic measurements, for example on air bubbles trapped in ice core samples, show that atmospheric carbon dioxide levels were never higher than 300 ppm over the past one million years (Lüthi et al. 2008). In most calculations about global warming, the global average for atmospheric carbon dioxide prior to the Industrial Revolution is taken as 280 ppm. As shown in Fig. 8.6, atmospheric carbon dioxide reached a record high of ~416 ppm in early 2020. As explained above, there is overwhelming evidence that this increase is due to the burning of huge quantities of fossil fuels by humans over the last 150 years. These fuels have become indispensible for industrial use, land transport, motorised shipping, aviation, and daily lifestyle activities. Other human contributions to atmospheric greenhouse gases come from agricultural releases and the increasing release of synthetic green-house chemicals, such as hydrofluorocarbons. Additional contributing factors are the increasing numbers of ruminant farm animals, which release methane, and the large-scale loss of forested land (deforestation), which has decreased the global capacity of plants to absorb atmospheric carbon dioxide. The biomass of livestock animals is now far greater than the biomass of wild mammals and birds (Bar-On et al. 2018). The relative contributions of different sectors of the economy to greenhouse gas emissions are illustrated in Fig. 8.10. If our present way of life does not change, it is projected that atmospheric carbon dioxide could reach 900 ppm by the end of the twenty-first century. This would result in an estimated further increase to global average temperatures of between +2.6 °C and +4.8 °C (IPCC 2014).

The effect of the Covid-19 pandemic on countries around the world in 2020 has also provided a sharp view on how current human activities affect the atmosphere. With personal travel restrictions, vastly decreased air travel and lock-down isolation taking place simultaneously in many countries during the first half of 2020, climatologists recorded an almost instantaneous decrease in atmospheric nitrogen dioxide (a greenhouse gas produced mostly from vehicle emissions) and the quantity of small particles in the air. As an example of the rapidity of this decrease, the image from NASA's satellites for atmospheric nitrogen dioxide over the southeastern USA in March–April 2020 is compared to the five-year average for the same month (Fig. 8.11).

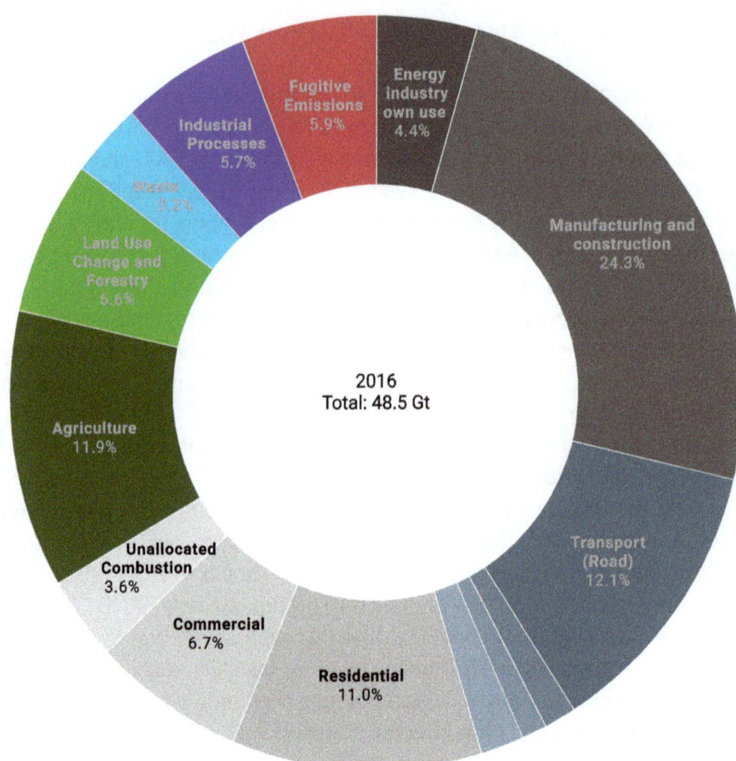

Fig. 8.10 Greenhouse gas emissions by economic sector, 2016 data. Percentages are calculated from estimated global emissions of all Kyoto greenhouse gases, converted to CO_2 equivalents (From Earthcharts.org [online resource, CC-BY], from Wikipedia [2020], English version)

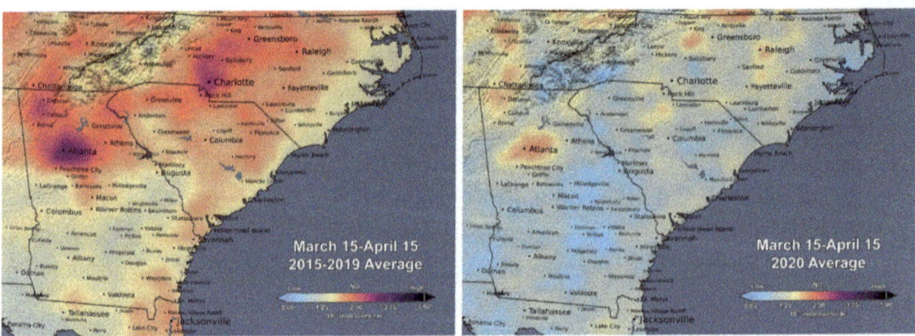

Fig. 8.11 Atmospheric nitrogen dioxide levels over the south-eastern USA, from NASA's satellite images, comparing the average for March–April 2015–2019 with March–April 2020 (From https://svs.gsfc.nasa.gov/4810. *Credit* NASA's Scientific Visualization Studio)

Similar air quality improvements were reported at different sites around the world, including for the major industrial areas in China and India. Data from Europe on causes of death showed that deaths due to air pollution were strongly reduced (Myllyvirta and Theriot 2020). The massive decrease in use of land transport and industrial activities during the first lockdown period in 2020 was also manifested as a striking decrease in seismic noise, i.e., vibrations in the Earth's crust (Gibney 2020; Lecocq et al. 2020). The first Covid-19 lockdown period also led to a remarkable 17% decrease in daily global carbon dioxide emissions in early April 2020 (Le Quéré et al. 2020) and an overall 8.8% decrease in these emissions in the first half of 2020 (Liu et al. 2020). However, the temporary nature of lockdown in each country means that there was no rapid effect on the mean global atmospheric carbon dioxide level (https://www.esrl.noaa.gov/gmd/ccgg/trends/). The effects of the global response to the pandemic on climate are very likely to be transient. Nevertheless, these very dramatic changes, that are clearly due to the global alternation in human activities, provide a preview of what could be achieved if concerted new policies and actions to reduce fossil fuel energy consumption and greenhouse gas emissions could be set up around the world (discussed further in Sect. 8.6).

8.4 Impacts and Consequences for Ecosystems and Biodiversity

The increases in average global temperatures and sea-levels that we have outlined in Sect. 8.1, in conjunction with increased atmospheric carbon dioxide and other greenhouse gases (Sect. 8.2), have many follow-on consequences for the global environment and the habitats of living organisms. This is particularly apparent for the Arctic and Antarctic regions, which because of their latitudes are warming almost twice as fast as other regions, resulting in the melting of ice and destruction of polar habitats. International attention has focused on polar bears in the Arctic, that depend on sea-ice for living, breeding, travel and access to their main prey, seals. As sea ice decreases, more polar bears are being forced onto land to find other food sources. This brings them into conflict with humans. Polar bears evolved to survive in the extreme cold of the Arctic. Warmer summer temperatures put stress on their physiology and ability to reproduce successfully. Bears are thinner and their populations are decreasing in some regions (https://arcticwwf.org/newsroom/stories/polar-bear-assessment-brings-good-and-troubling-news/). Similar issues face other large Arctic animals, including walruses and caribou. In contrast, as the Arctic tundra has warmed, certain trees and shrubs have started to grow further north, which provides an extended range of habitat for animals such as snowshoe hares (Tape et al. 2016).

Challenges to the normal physiology of animals and plants are also taking place in equatorial regions where desert regions are becoming hotter and more extensive through a combination of global warming, reduced rainfall, and increased human activities. For example, the Sahara desert became ~10% larger during the twentieth century (Thomas and Nigam 2018). Almost 30% of current dry land regions are predicted to break down as ecosystems and/or become uninhabitable if global warming continues unchecked (Berdugo et al. 2020). Land is also being lost or transformed due to the rising sea level (Fig. 8.3). Coastal flooding is increasing, causing changes in biodiversity as freshwater environments are transformed to estuarine. For example, it is predicted that Bangladesh may lose 11% of its land by 2050. Low-lying islands are also shrinking in size. The loss of land is particularly prominent for the low-lying South Pacific islands, which are the first nations to be under direct threat from increased sea levels. As the sea level rises, the environments of land plants and animals are constricted and land for human habitation and agriculture is diminished. Frequent flooding also increases the likelihood of infectious diseases of humans (Bakare et al. 2020). It is projected that most South Pacific atoll islands will become uninhabitable by the mid twenty-first century (Storlazzi et al. 2018). Global warming is also linked to an increased likelihood of extreme weather events such storms, heat-waves, or droughts and may be linked to increasingly severe cyclones or hurricanes (Fig. 8.12). It is not clear yet whether global warming itself has affected the frequency of these storms, but these severe weather events pose direct threats

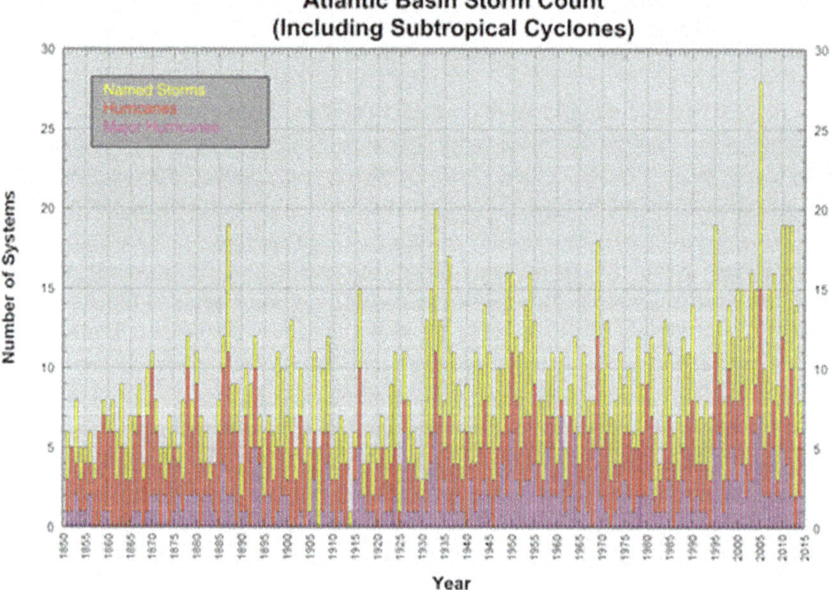

Fig. 8.12 The occurrence of Atlantic hurricanes between 1850 and 2015. The incidence of major hurricanes increased in the twentieth century (Data from NOAA [https://www.nhc.noaa.gov/climo/])

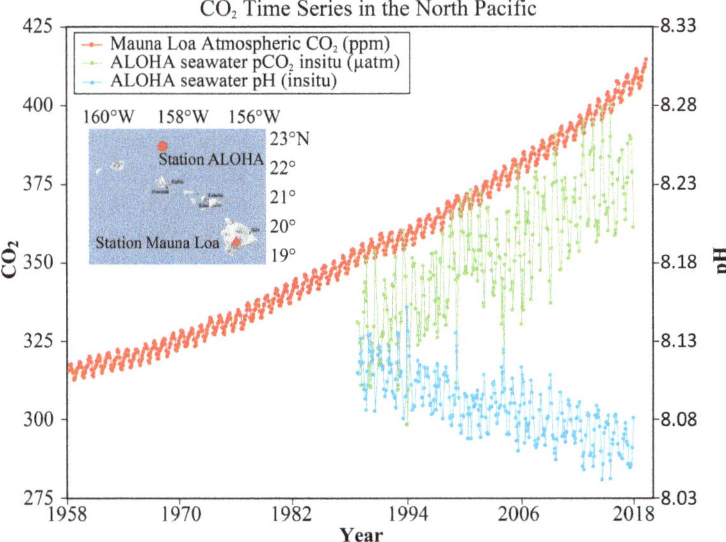

Fig. 8.13 The decrease in pH of the North Pacific Ocean between 1988–2018 (Data from NOAA [https://pmel.noaa.gov/co2/story/QualityofpHMeasurementsintheNODCDataArchives]. *Data* Mauna Loa (ftp://aftp.cmdl.noaa.gov/products/trends/co2/co2_mm_mlo.txt) ALOHA (http://hahana.soest.hawaii.edu/hot/products/HOT_surface_C02.txt). Ref: J.E. Dore et al., 2009. Physical and biogeochemical modulation of ocean acidification in the central North Pacific. *Proc Natl Acad Sci USA* **106**:12235–12240)

to life especially in low-lying islands and coastal regions (Stillman 2019). Preparations are also being made in Europe to adapt to more frequent severe weather events (EASAC 2018).

One of the most fundamental consequences of ocean warming is ocean acidification. Because around 30% of atmospheric carbon dioxide is absorbed by the oceans, the continuing increase in atmospheric carbon dioxide (Fig. 8.6), means that more carbon dioxide has become dissolved in the oceans to produce the weak acid, carbonic acid. As a consequence, over the last thirty years, ocean pH has decreased by 0.1 pH unit (Fig. 8.13). This may sound like a small change, but because pH is measured on a logarithmic scale, 0.1 pH unit signifies a ~30% increase in ocean acidity.

In addition to direct effects of a more acidic pH on living organisms, the change in ocean pH has major consequences for carbonate chemistry, because fewer free carbonate ions are available under the more acidic conditions. This change to ocean chemistry is distinct from any event over the last 300 million years (Hönisch et al. 2012) and is exerting large-scale effects on marine organisms that have shells or exoskeletons built from calcium carbonate (Jiang et al. 2019). Compared to samples collected in the nineteenth century by the *HMS Challenger* expedition, modern foraminifera (single-celled marine plankton with an outer calcium carbonate shell) collected from the same sites in the Pacific Ocean have shells that are 30–76% thinner, depending on the species (Fox et al. 2020). Under laboratory experiments of exposure to reduced pH, representative species formed thinner calcium carbonate shells and/or put more energy into shell-building (reviewed by Gazeau et al. 2013). In the oceans, the weakening of these organisms has adverse effects on their survival and so also affects food chains. This threatens the viability of other invertebrates, fish, birds and other ocean animals. Coral reefs are also very vulnerable to damage through loss of calcium carbonate. This effect is compounded by their susceptibility to increased ocean temperature, which triggers a stress response termed coral-bleaching. As well as threatening the viability of the corals, these processes reduce the habitats of the many organisms that live in coral reefs (Spalding and Brown 2015).

A further major issue is that certain massive changes in the natural environment, such as the diminishment of ice sheets in the Arctic or Antarctic, or large scale ($>20\%$) deforestation of the tropical rainforests, are predicted to act as "tipping points" that would destabilize the global climate or raise the sea level in major, irreversible ways that could have additional, unforeseen consequences. Once started, these processes could have cascading effects leading to situations that humans would not know how to deal with and, at worst, a catastrophic, global crisis point for life on Earth (Lenton et al. 2019). In previous climate change models, it was thought that 5 °C of global warming would activate tipping points, but recent models indicate that tipping points could be activated with less than 2 °C of warming (IPCC 2019).

As natural ecosystems come under unprecedented and accelerating environmental challenges, other aspects of human activities are also having negative effects on the natural world and provide additional stresses. Human land use has altered the majority of the land and freshwater environments and has fragmented natural landscapes and ecosystems in ways that make them less resilient to further changes. In 2016, it was calculated that only 23% of the land remains as wilderness (Watson et al. 2016) and a later report also estimates that human activities have severely altered 75% of the land area on Earth (IPBES 2019). Human activities have (either accidently or deliberately) moved many species around the world, with negative consequences on indigenous species. Over-fishing has affected ocean ecosystems and led to a decline of many fish species. Although nuclear power is seen as "clean" energy, major disasters such as at Chernobyl (1986) or Fukushima Daiichi (2011) had major deleterious environmental consequences. Artificial light pollution is affecting the ability of species to breed, particularly in coastal regions (Ayalon et al. 2020), and noise pollution also has negative effects on wild-life (Shannon et al. 2016). Human actions are putting many species under stress from fire events (Kelly et al. 2020). Pollution of the environment with chemicals, pesticides, and toxic waste products is having devastating effects on the populations of many species. For example, an analysis of 529 species of North American birds based on 50 years of population monitoring surveys indicated a net loss of 2.9 billion birds over this period, leaving, overall, 29% of the numbers of birds compared to 1970 (Rosenberg et al. 2019). Sharp decreases in pollinating insects due to extensive use of neonicotinoids have led to declining populations of insectivorous birds (Hallmann et al. 2014). Climate change itself is also a likely cause of decreasing bumblebee populations (Soroye et al. 2020). Plastic waste including microparticles has contaminated all ecosystems and the bodies of living organisms;

we do not yet understand the issues for any species in living with these particles over generations (Cox et al. 2019; Borrelle et al. 2020).

The World Wildlife Fund Living Planet report of 2018 reported that the population sizes of more than 4,000 species of animals had shrunk on average by 60% since 1970, with even more dramatic decreases occurring in tropical regions and freshwater environments (World Wildlife Fund 2018). The report of 2020 contains even bleaker data that populations have now shrunk by an average 68%, mainly due to habitat degradation and loss (World Wildlife Fund 2020). A massive landmark report by the Intergovernmental Science-Policy Platform on Biodiversity and Ecosystem services (IPBES), on state of biodiversity on Earth has estimated that 1 million of the known 8–9 million species on Earth are in danger of becoming extinct in the next fifty years. These include over 40% of amphibian species, 33% of reef-forming corals and over 30% of marine mammals (IPBES 2019). An assessment of extinctions of vertebrate species since 1500 showed that the extinction rate since then has been one hundred times higher than the "background" (natural) rate of extinction, and that most of the extinctions occurred after 1900 (Ceballos et al. 2015). At least 515 species of land vertebrate are on the brink of extinction, because their populations have diminished to less than 1,000 individuals (Ceballos et al. 2020). From survey data on 538 plant and animal species, of which 256 species had already suffered local extinctions, it has been calculated that 16–30% of all the examined species could be entirely extinct by 2070 (Román-Palacios and Wiens 2020). In comparison to this focus on animals and green plants, we know very little about the effects of pollution, global warming and the related environmental changes on microorganisms, yet these underpin all life, along with soil quality and biodiversity, and will have vital roles in any regeneration of ecosystems (Cavicchioli et al. 2019; FAO, ITPS, GSBI, SCBD and EC Report 2020).

In the face of the massive, ongoing destruction of ecosystems and species loss, the Convention of Biological Diversity, a part of the UN Environment Programme, set up a strategic plan for the years 2011–2020 known as the Aichi Biodiversity targets (https://www.cbd.int/sp/targets/). Whilst this plan has entirely laudable goals (for example, strategic goal A is to "Address the underlying causes of biodiversity loss by mainstreaming biodiversity across government and society"), unfortunately, by 2020, progress was far behind what was envisioned in 2011. For example, Target 5 states:

"By 2020, the rate of loss of all natural habitats, including forests, is at least halved and where feasible brought close to zero, and degradation and fragmentation is significantly reduced". Very sadly, globally, we are nowhere near meeting this goal or other Aichi targets. A specific national case-study is provided by an independent report on the UK's progress by the Royal Society for the Protection of Birds charity, which concluded that 17 out of the 20 targets have not been met (RSPB 2020). Similarly, the UN Global Diversity Report 5 (2020) concludes: "At the global level none of the 20 targets have been fully achieved, though six targets have been partially achieved".

A movement to set up an international legal framework to protect environments, the "Stop Ecocide" campaign and Foundation, was founded in 2017 in the UK by the lawyer Polly Higgins and environmental activist Jojo Mehta. Ecocide is defined as the severe damage or destruction of the natural world and Earth's ecosystems by human activity. The Foundation seeks to draw up a legal definition of ecocide and make ecocide an internationally recognised crime against humanity, so that companies, states, or individuals who engaged in large-scale, damaging environmental actions would be subject to criminal prosecution. Other efforts to legal recourse are the increasing number of cases brought, often by young people, to hold governments to account over climate change commitments and their duty to preserve natural public resources for future generations. Hundreds of cases have been brought around the world and there have been some successes, such as the block by the UK Court of Appeals to a third runway at Heathrow airport (Viglione 2020). In February 2021, the French government was convicted by a Paris court of failing to keep its promises and targets to reduce greenhouse gas emissions, in a case brought by four French environmental organisations. The ruling also included that compensation in kind for 'ecological damage" was admissible.

8.5 Effects of the Increasing Human Population

The likelihood of success for efforts to reverse or compensate for the global climate change that humans have unfortunately set in motion will be affected by the further growth of the human population. In the nineteenth century, the world population increased gradually with the start of improvements to general living conditions in some countries. With improvements to health (Sect. 6.1), diet and reduced child mortality, the global population then increased much more rapidly throughout the twentieth century (Fig. 8.14). Since 1950, the population has more than tripled to over 7.7 billion people in 2019 (UN World Population Prospects 2019).

The human population is projected to grow to around ten billion people by 2050 (Fig. 8.14). The highest rate of population growth was in the 1960s, and the overall rate of growth has been decreasing consistently over the last twenty years (purple line in Fig. 8.14). At present, the global population is increasing by 80 million people a year. This overall trend includes considerable differences in population dynamics in different countries. In many westernised countries, populations are static or falling due to small family sizes. More than half the projected growth in the global population up to 2050 is expected to take place in sub-Saharan Africa. Increasing lifespan will also play a part, with the global average life expectancy projected to increase from 72.6 years in 2019 to 77.1 years by 2050. The decade 2010–2020 has seen considerable movements of people as emigrants or refugees, and it is expected that large-scale migration will continue to impact on population growth in different regions of the world (UN World Population Prospects 2019). An analysis by the Institute for Economics and Peace (IEP) projects that over one billion people may become migrants by 2050, due to an interplay of local population growth, ecological crises, natural disasters and civil unrest or conflicts (IEP Ecological Threat Register 2020).

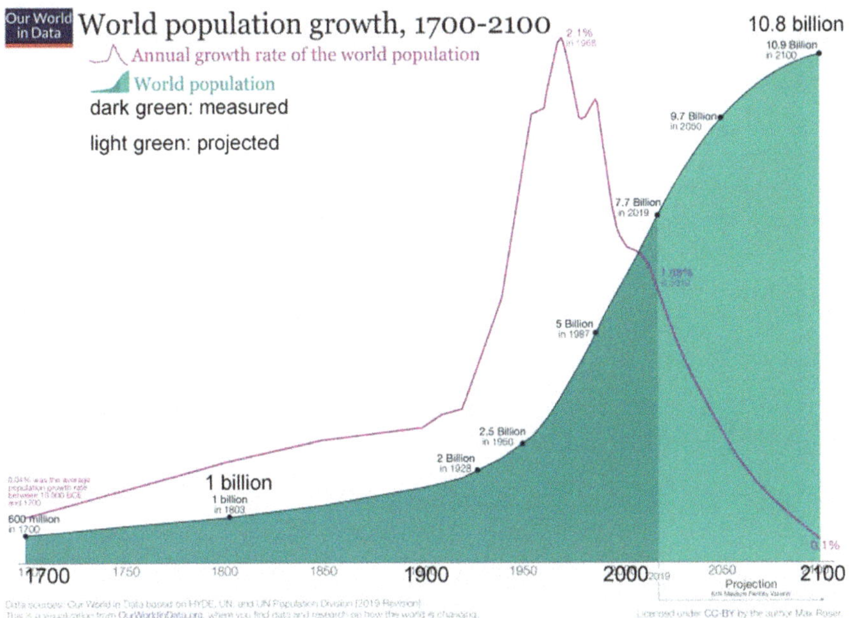

Fig. 8.14 World population growth from 1700. Growth after 2019 is projected according to the UN medium-fertility projection model (From Max Roser, Hannah Ritchie and Esteban Ortiz-Ospina (2013)—"World Population Growth". Published online at OurWorldInData.org. Adapted from: 'https://ourworldindata.org/world-population-growth' [Online Resource, CC-BY])

The vast increase in the number of humans on the Earth over the last seventy years along with economic systems that focus on high consumption lifestyles have put huge stresses on the natural world. These effects have consequences that feedback and interact with the processes of global warming and climate change. More people take up more space on the land and need more food and space for housing and transport links. These demands reduce the amount of land available for natural ecosystems, brings humans into closer proximity with wild animals and facilitates transfer of zoonotic diseases. Human activities have degraded wilderness environments, rivers, lakes and the oceans. For example, the area of mangrove swamps in Myanmar decreased by 63% between 1996 and 2016 as agricultural areas expanded (De Alban et al. 2020). A global study by satellite imaging revealed that land wilderness areas the size of Mexico were lost between 2000 and 2013 (Williams et al. 2020). Over-exploitation of natural resources such as forests and oceans is affecting the resilience and sustainability of ecosystems and the biosphere to such an extent that a business analysis of Biodiversity and Ecosystems Services by the Swiss insurance company Swiss Re determined that 20% of countries have ecosystems at risk of collapse (Schelske et al. 2020). The enormous global increase in use of fossil fuel energy since 1950 (Fig. 8.15), has increased the release of greenhouse gases and pollutants. Increasingly intensive agricultural practices are depleting soil quality and

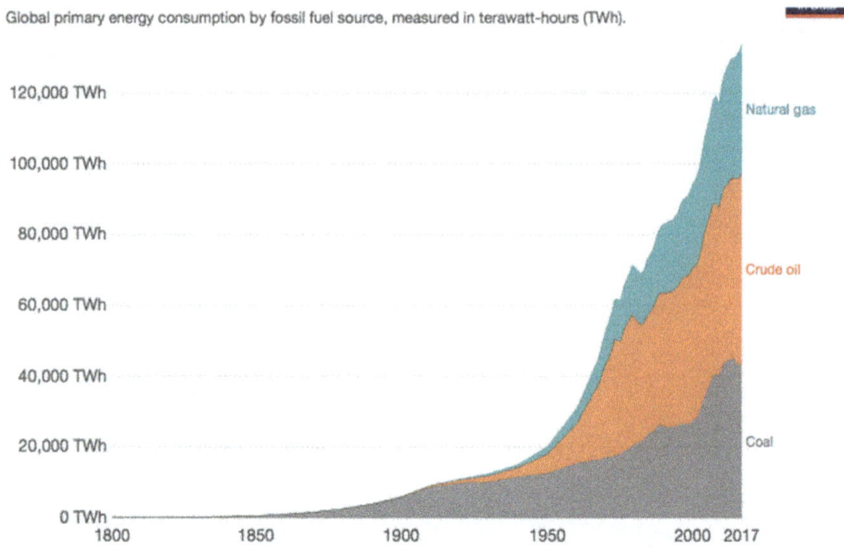

Fig. 8.15 The extreme growth in global consumption of fossil fuels over the twentieth century (From Hannah Ritchie and Max Roser [2017], "Fossil Fuels". Published online at OurWorldInDa ta.org. Retrieved from: 'https://ourworldindata.org/fossil-fuels' [Online Resource, CC-BY]. *Source* Vaclav Smil (2017). Energy Transitions: Global and National Perspective & BP Statistical Review of World Energy OurWorldlnData.org/fossil-fuels/ · CC BY)

exacerbating many effects of climate change. For example, the preference for meat-based diets in many countries has led to vastly increased populations of ruminant farm animals that produce large amounts of the greenhouse gas, methane.

The scale of environmental changes brought about by humans is so great that a working party of the International Commission on Stratigraphy voted by a majority in 2019 to designate a new geological epoch, the Anthropocene (Subramanian 2019). The combination of on-going global warming and the environmental changes it brings, along with the demands on natural resources of maintaining the modern human lifestyle for billions of people, is proving toxic to nature in numerous ways. Human population growth and consumption must both be addressed if these effects are to be mitigated (Ganivet 2019).

8.6   Efforts to Mitigate Against Climate Change

Human awareness of global warming and climate change as an issue that threatens the whole planet grew out of the environmental movement of the mid-twentieth century. From initial actions such as Earth Day that was begun in the 1970s to draw attention to environmental problems (earthday.org), progress has been made in various ways, which we discuss below. Nevertheless, the goal of controlling future global warming and reducing carbon emissions remains extraordinarily difficult. It would be extremely useful to define a quantitative relation between the amount of emitted greenhouse gases and the resulting temperature increase. However, the Earth is a very difficult system to study because it exchanges energy in many ways and produces heat by itself. For a body of the size of the Earth, a direct experimental determination of the effects of greenhouse gas on heat absorption is not possible. Therefore, computational models have been developed to model future climate change, based on assumptions of different greenhouse gas inputs. The quality of these models can be examined by determining how well they simulate known past or current climate change events.

The original industrialised countries have emitted, and continue to emit, a very high proportion of the total global greenhouse gas emissions, due to their high consumption lifestyles and political emphasis on economic growth. Efforts to reduce consumption by educating people to change their lifestyles have continued for many years and individual choices certainly have a role to play. However, the finite nature of fossil fuel resources and the pace at which extensive and global changes need to be made to keep global warming within a 1.5 °C to 2 °C target range (see Sect. 8.7 below for discussion of the 2015 Paris Agreement), mean that changes in lifestyles and societal values are needed as soon as possible across the human population. Crucially, moving to low- or zero-carbon, green economic models will depend on renewed appreciation of the essential roles of the wild natural world, new low- or zero-carbon technologies, and changes in business and agricultural practices, along with effective governmental policies and binding international agreements. For example, it has been modeled that current global food system emissions alone could result in the global warming targets being missed (Clark et al. 2020).

Over the last decade, a number of countries have started to sharply decrease their reliance on coal to generate the electricity supply. As well as the need to switch to new fuel sources as coal supplies have become used up, this move has been driven by governmental initiatives that encouraged adoption of renewable technologies such as solar, wind and wave power. Investment in innovations for these technologies has further stimulated their uptake: for example the efficiency of solar panels has increased whilst their cost has dropped sharply. The cost of batteries for energy storage is also decreasing. Taking the UK as an example, in 2019 119TWh of renewable energy was generated (mostly from biofuels and wind power), and this accounted for 36.9% of the electricity generated (https://assets.publishing.service.gov.uk/government/uploads/system/uploads/attachment_data/file/875410/Renewables_Q4_2019.pdf). In the third quarter of 2019, a landmark event was reached when renewable energy sources in the UK provided more energy than fossil fuels for the first time since 1882. In May and June 2020, coal-free electrical power generation was achieved for a 67-day period for the first time (https://www.nationalgrideso.com/news/great-britains-new-record-longest-period-coal-free-electricity-generation). During this time, UK power generation came from renewable sources (35%), gas (32%), nuclear (21%) and imports (11%).

Although technological innovations are helping to reduce reliance on fossil fuels and can be drivers of new economic models, it would be naïve to think that technology could replace nature. The complex web of life involves millions of biological species that have evolved over many billions of years. We do not yet know all the species on the modern Earth, let alone the real consequences of enforcing rapid changes on ecosystems though irreversible species loss. One example of current ideas on how technology could assist nature is the idea to use tiny, robot flying machines as plant pollinators to supplement bees as pollinators. While bee-like robotic pollinators might be feasible over a small area, the machines would not have all the characteristics of real bees and, for the plants, would not replace the complex network of plant-pollinator relationships of natural pollinators. The manufacture of the vast numbers of robots that would be needed would have its own environmental impact and would produce additional emissions (Potts et al. 2018). It makes much more sense to support the natural insects that exist now, which carry out wide scale pollination at no cost, and identify ways to increase the populations of wild bees and other pollinating insects.

While renewable energy is a very important step towards decreasing carbon emissions, the generation of "clean" energy needs to be coupled with reductions in dependence on carbon and an overall reduction in greenhouse gas emissions in all areas of human life. New technologies can provide energy solutions but may have their own challenges to be sustainable: for example, battery technology relies on access to necessary metals and minerals. Transport is another major source of emissions (Fig. 8.10), and technological advances for electric- or hydrogen-powered cars, shipping and even electrically powered airplanes are in progress. Delivery of small goods by drone can reduce road transport. New styles of house-design can produce houses that require minimal heating or cooling, and older houses can be retro-fitted to improve insulation and the method of heating. If more people worked from home, cities could be made more pleasant to live in and commuter traffic reduced. Reducing reliance on single-use plastic packaging alongside innovations to increase the economic value of plastic recycling can also reduce industrial and household consumption of oil-based goods. The identification of a bacterial enzyme capable of degrading plastic has opened up new prospects for handling plastic waste (Austin et al. 2018). Many other encouraging examples are reported frequently in the news. There are also moves to reduce extraction of fossil fuels and leave the remaining coal and oil reserves in the ground. Since agriculture is also a major source of emissions (Fig. 8.10), changes to agricultural practices and food supply systems will also be essential (Clark et al. 2020).

A parallel strategy against global warming is to remove carbon dioxide from the atmosphere by so-called negative emissions technologies. A technologically based solution is the concept of artificially capturing carbon dioxide from the air and concentrating it for storage underground or for recycling by industrial processes into synthetic fuel. The possibility of capturing and storing carbon dioxide in alkaline rock waste from mining industries is also under investigation (Service 2020). Since plants, especially trees, take up carbon dioxide as a natural part of photosynthesis, another approach is to restore forests and plant extra trees on a massive scale. Forests have been estimated to store about 400 Gigatonnes of carbon (Pan et al. 2013). An IPCC report has estimated that negative emission technology would need to have the capacity to take up at least 0.5–3 Gigatonnes carbon per year (or, in the worst case climate change scenario, a capacity of 7–11 Gigatonnes carbon per year), to mitigate the current climate change trajectory (IPCC 2018). Tree-planting projects are currently underway at various scales; expanding this natural capacity for carbon dioxide uptake can also potentially give more time for societies to develop technological carbon-capture solutions and adapt to a low/zero carbon emission way of life. Some of the large-scale goals that have been set up include planting a tree for every person in the UK by 2025 (Woodland Trust) and the massive planting of tree seedlings in the "Green Legacy" project in Ethiopia (https://pmo.gov.et/greenl egacy/).

A study that used machine learning to map the global potential for tree coverage calculated that, under current climate conditions, an extra 0.9 billion hectares of canopy cover could theoretically be planted globally that could store a total of 205 Gigatonnes of carbon. This project would involve planting an estimated 517 billion trees (Bastin et al. 2019). The study raised many questions (see discussion at https://climate.nasa.gov/news/2927/examining-the-viability-of-planting-trees-to-help-mitigate-climate-change/) such as how quickly trees can be planted, the types of trees that will best support wildlife, or whether there would be unintended consequences on availability of water for other organisms, or other issues. The types of trees to be planted is an important consideration in relation to the carbon storage potential, as faster-growing trees typically have shorter lifespans (Brienen et al. 2020). We also need to learn more about how different tree species cope with increased average temperatures and more frequent droughts and storms (Brodribb et al. 2020). However, there is no doubt that increasing tree cover offers an immediately accessible route to decreasing atmospheric carbon dioxide which would benefit other organisms as well as humans. Other routes to increasing natural carbon capture would be to adopt sustainable agricultural practices on a large scale to improve both soil quality and carbon sequestration (UN Environmental Program Foresight Brief 2019). Aquatic coastal ecosystems with abundant mangroves, seagrass, or underwater kelp forests are also being destroyed rapidly, yet can sequester carbon even more effectively than forests (Macreadie et al. 2019). Repairing and supporting these habitats is another route that could increase natural carbon sequestration (https://www.thebluecarbonin itiative.org). An analysis of past conservation efforts concluded that recovery of healthy oceans could be achieved by 2050 if only the pressures of climate change can be mitigated (Duarte et al. 2020).

8.7 International Conventions on Climate Change and Future Projections

It is clear that many solutions must be pursued in parallel if the ongoing increase in atmospheric carbon dioxide is to be halted and then returned closer to historic levels. This gigantic global problem will necessitate effective international agreements and coordinated actions. This area has been pursued over the last forty years by international bodies set up specifically to address climate change and as a component of the activities of the United Nations. The first international scientific conference to discuss climate change and its implications was the World Climate Conference, organised by the World Meteorological Organization (WMO), which took place in Geneva in 1979. This conference later led to the set up of the Intergovernmental Panel on Climate Change (IPCC) by the WMO and United Nations Environment Programme in 1988. Initially, these conferences focused mostly on the accumulating scientific data that climate change is taking place, but over time they have raised international awareness of the issue. As the scientific data have become compelling, the conferences have become forums for governmental policy-making and international agreements.

The IPCC has a mandate to "assess the state of existing knowledge about the climate system and climate change; the environmental, economic, and social impacts of climate change; and the possible response strategies" (ipcc.ch). Its first assessment report was released in 1990 and, in addition to confirming the scientific evidence for climate change, provided policy-makers with up-to-date, accurate information. This led to the start of UN negotiations for an international treaty by which countries would recognise climate change and begin to develop strategies to counteract it. The resulting UN Framework Convention on Climate Change was signed by 154 states and the European Commission in 1992 at the Earth Summit in Rio de Janeiro and was ratified in 1994. Its goal has been to "stabilise greenhouse gas concentrations at a level that would prevent dangerous anthropogenic interference with the climate system". The "Conference of the Parties (COP)", has continued to meet regularly since 1995, to discuss commitments, mechanisms, procedures and support for developing countries. Goals to address climate change and the related degradation of the natural environment are also included in the seventeen Sustainable Development Goals of the UN (https://www.un.org/sustainabledevelopment/sustainable-dev elopment-goals/).

As part of its activities, the IPCC produces reports that assess climate change knowledge and model potential future scenarios. In the IPCC 2014 report, four "Representative Concentration Pathways" (RCP) were adopted for modelling the possible trajectories and extent of increasing atmospheric greenhouse gas concentrations over the twenty-first century. The models are presented in terms of radiative forcing (the difference between solar radiation absorbed by the Earth and energy emitted into space) with units of Watts per m^2. The models range from RCP2.6, in which mitigation efforts are *likely* to keep the overall temperature increase below $+2\ ^\circ$C, through intermediate scenarios, RCP4.5 and RCP6.0, and a high emissions scenario, RCP8.5. Continuing with no change to present emissions is predicted to result in a high RCP, between RCP6.0 and RCP8.5 (IPCC Synthesis Report 2014; https://ar5-syr.ipcc.ch/topic_futurechanges.php).

COP 21 was held in Paris in 2015. By this time, the urgency of restricting greenhouse gas emissions was generally recognised. The Paris Agreement from COP 21 was the first legally binding, global climate agreement. It was signed by 195 countries in 2015 and ratified in 2016. The agreement specified for countries to take steps to reduce their greenhouse gas emissions with the goal of *limiting future global temperature increases to less than 2 °C above pre-industrial levels and also to pursue efforts to restrict any temperature rise to 1.5 °C*. The signatory countries also agreed to strengthen national arrangements to deal with the impacts of climate change and to increase international support to developing countries for adaptations to reduce emissions and/or deal with ongoing effects of climate change. By 2019, 183 countries had submitted nationally determined contribution plans for their individual actions.

The IPCC special report of 2018 (IPCC 2018) emphasised the huge additional dangers to the Earth and its living beings if the average global temperature was to increase by 2 °C instead of ~1.5 °C. Given that the global average temperature has already risen by $+1.16$ °C (Sect. 8.1), an important communication in this report was the estimate that a 1.5 °C rise in temperature could be reached *within 11 years* unless global carbon dioxide emissions can be cut drastically (IPCC 2018). Although debate continues over the basis of the emissions calculations, the sombre conclusion remains that it is extremely urgent to take global actions to change the human lifestyle. As stated in an analysis report on the IPCC data: "No matter what carbon budget is used, there is still less than 0.5°C additional warming to go before 1.5°C is passed and only a few decades before the world has to reach net-zero – and then net-negative – emissions" (Carbon Budgets 2018). Overall, the likelihood of keeping to the Paris 2015 goals is diminishing with time.

The COP conferences have made only limited progress since 2015. Since 2018, public demonstrations for governmental initiatives to swiftly reduce greenhouse gas emissions and counteract climate change, especially by young people, have become increasingly prominent around the world. The young Swedish climate activist Greta Thunberg began a personal climate strike action in summer 2018 by sitting outside the Swedish Parliament and has rapidly become the figurehead of a global movement. Environmental protests were held in many countries during 2019. When Greta Thunberg addressed the United Nations in New York on 23 Sept 2019, she emphasised "With today's emissions levels, that remaining CO_2 budget will be entirely gone within less than eight and a half years". During 2019, many governments, regions, cities and diverse organisations took steps in declaring "Climate Emergencies" to accelerate their own efforts to reach net-zero emissions. For many countries, the current target date to reach net-zero carbon emissions is 2050. However, the UN Environment Programme's emissions gap report of 2019 estimates that, even if current agreed targets are met, global emissions will still continue to grow. The report estimates that, by 2030, emissions will be 27% and 38% higher than are required to limit global warming to 2 °C or 1.5 °C, respectively. The scale of the task in the next decade is emphasised by the fact that emissions in 2030 would need to be either

25% or 55% lower than in 2018 to hold the temperature increase to either a 2 °C or 1.5 °C increase, respectively (https://www.unenvironment.org/resources/emissions-gap-report-2019).

Given that global warming is still less than +1.5 °C, it may seem that there is good scope to restrain global warming to +1.5 °C. However, it is important to be aware that the +1.5 °C figure is for *global average* warming: the land warms faster than the oceans and large cities have their own micro-climates. Warming in cities, high mountain regions and at the poles is already at a higher level than the global average. We discussed some of the positive feedback effects from ice-melt or methane release in Sect. 8.1; the activation of these and other feedback effects means that even if human society completely stopped its greenhouse gas emissions today, warming of the atmosphere and the rising of sea levels would not cease for years.

Overall, if human society does not take effective action very soon, climate change and destruction of nature will bring huge additional risks to humans. We need to recognize that, although nature would continue without us, humans cannot survive without the natural world. Loss of the necessities of life are the most central and prominent risks from continuing climate change: billions of people will face reduced water quality and water supply, and/or reduced food crops due to insufficient pollination and desertification of land (Chaplin-Kramer et al. 2019). Coastal or low-lying communities will need to be abandoned, reinvented, or relocated as severe storms and flooding increase. Because human populations are concentrated on land, and most of the projected growth in the human population (Sect. 8.5) will occur in tropical regions, it is estimated that the average temperature increase experienced by humans by 2070 could be around +7.5 °C (Xu et al. 2020). Equatorial regions or cities would become uninhabitable due to the high temperatures, water shortages and desertification. Human physiology cannot tolerate external temperatures above 35 °C for extended periods of time, and so heat-related deaths are predicted to soar (Sherwood and Huber 2010). Indeed, a model of the effects of further global warming on mortality based on 451 locations in 23 countries with different climates has projected that warming *above 2 °C* would be sufficient to increase heat-related deaths (Vicedo-Cabrera et al. 2018). Collectively, these changes would result in large-scale migrations of people away from equatorial regions, along with increased poverty and inequality. With regard to economic factors, it has been modeled that the financial cost of ignoring climate change will be 4–5 times higher than the cost of the measures needed to hold to an average temperature increase of +1.5 °C (Hoegh-Guldberg et al. 2019).

Unfortunately COP25, held in December 2019 in Madrid, although the longest ever COP conference, did not reach clear outcomes on new steps to restrain further global warming. Much will depend on COP26, to be held in Glasgow, UK (https://uk.cop26.org). This conference was postponed from 2020 to November 2021 due to the covid-19 pandemic (https://sdg.iisd.org/events/2020-un-climate-change-conference-unfccc-cop-26). In contrast, smaller groupings such as regions, cities or institutions are making clearer progress and have ambitious goals. In 2015, the nation of Wales passed into law the Well-being of Future Generations (Wales) Act 2015. This was the first time that regenerative and sustainability practices were put into law at a national level. The Act crucially states "sustainable development… means that needs of the present are met without compromising the ability of future generations to meet their own needs" (https://futuregenerations.wales/wp-content/uploads/2017/02/150 623-guide-to-the-fg-act-en.pdf). In 2019, the city of Bristol was the first city in the UK to set a goal to be net carbon neutral by 2030, the One City Climate Strategy (https://www.bristol.gov.uk/policies-plans-strategies/council-action-on-climate-change). In Denmark, the city of Copenhagen is working to be carbon-neutral by 2025 (CPH 2025 Climate Plan 2011). Many of the world's largest cities have entered a network, C40 Cities, to address climate change and deliver the goals of the Paris Agreement. Over 50 of these cities are reported to be on-target to decrease their emissions from 2020 onwards (https://www.c40.org/about).

In response to the covid-19 pandemic, many cities are starting to accelerate infrastructural changes to improve air quality and to promote walking and cycling over car transport. Other developments are not so constructive: urgently needed societal efforts to address climate change have been retarded or abolished by the covid-19 pandemic (see Chap. 6). In many countries, many millions or billions of funding have been committed to support the economy and jobs throughout the pandemic. This is used as an argument not to pass climate laws that were already planned before the crisis. Political leaders who ignore the dangers of global warming continue to have a lot of support, yet, at the time of writing it appears that climate change with global heating has become the most pressing general threat to life on Earth.

References

Abram NJ, McGregor HV, Tierney JE, Evans MN, McKay NP, Kaufman DS (2016) Early onset of industrial-era warming across the oceans and continents. Nature 536(7617):411–418. https://doi.org/10.1038/nature19082

Allen RC (2017) The industrial revolution: a very short introduction. Oxford University Press, UK. ISBN 978-0-19-870678-6

ArchaeoGLOBE Project (2019) Archaeological assessment reveals Earth's early transformation through land use. Science 365:897–902. https://doi.org/10.1126/science.aax1192

Austin HP, Allen MD, Donohoe BS, Rorrer NA, Kearns FL, Silveira RL, Pollard BC, Dominick G, Duman R, El Omari K, Mykhaylyk V, Wagner A, Michener WE, Amore A, Skaf MS, Crowley MF, Thorne AW, Johnson CW, Lee Woodcock H, McGeehan JE, Beckham GT (2018) Characterization and engineering of a plastic-degrading aromatic polyesterase. PNAS USA 115:4350–E4357. https://doi.org/10.1073/pnas.1718804115

Ayalon I, Rosenberg Y, Benichou JIC, Campos CLD, Sayco SLG, Nada MAL, Baquiran JIP, Ligson CA, Avisar D, Conaco C, Kuechly HU, Kyba CCM, Cabaitan PC, Levy O (2020) Coral gametogenesis collapse under artificial light pollution. Curr Biol S0960-9822(20):31582-7. https://doi.org/10.1016/j.cub.2020.10.039

Bakare AG, Kour G, Akter M, Iji PA (2020) Impact of climate change on sustainable livestock production and existence of wildlife and marine species in the South Pacific island countries: a review. Int J Biometeorol. April 10. https://doi.org/10.1007/s00484-020-01902-3

Bar-On YM, Phillips R, Milo R (2018) The biomass distribution on Earth. Proc Natl Acad Sci U S A. 115:6506–6511. https://doi.org/10.1073/pnas.1711842115

Bastin JF, Finegold Y, Garcia C, Mollicone D, Rezende M, Routh D, Zohner CM, Crowther TW (2019) The global tree restoration potential. Science 365:76–79

Bell RE, Seroussi H (2020) History, mass loss, structure, and dynamic behavior of the Antarctic Ice Sheet. Science 367:1321–1325. https://doi.org/10.1126/science.aaz5489

Berdugo M, Delgado-Baquerizo M, Soliveres S, Hernández-Clemente R, Zhao Y, Gaitán JJ, Gross N, Saiz H, Maire V, Lehmann A, Rillig MC, Solé RV, Maestre FT (2020) Global ecosystem thresholds driven by aridity. Science 367:787–790. https://doi.org/10.1126/science.aay5958

Borrelle SB, Ringma J, Law KL, Monnahan CC, Lebreton L, McGivern A, Murphy E, Jambeck J, Leonard GH, Hilleary MA, Eriksen M, Possingham HP, De Frond H, Gerber LR, Polidoro B, Tahir A, Bernard M, Mallos N, Barnes M, Rochman CM (2020) Predicted growth in plastic waste exceeds efforts to mitigate plastic pollution. Science 369:1515–1518. https://doi.org/10.1126/science.aba3656

Bova S, Rosenthal Y, Liu Z et al (2021) Seasonal origin of the thermal maxima at the Holocene and the last interglacial. Nature 589:548–553. https://doi.org/10.1038/s41586-020-03155-x

Brienen RJW, Caldwell L, Duchesne L, Voelker S, Barichivich J, Baliva M, Ceccantini G, Di Filippo A, Helama S, Locosselli GM, Lopez L, Piovesan G, Schöngart J, Villalba R, Gloor E (2020) Forest carbon sink neutralized by pervasive growth-lifespan trade-offs. Nat Commun 11(1):4241. https://doi.org/10.1038/s41467-020-17966-z

Brodribb TJ, Powers J, Cochard H, Choat B (2020) Hanging by a thread? Forests and drought. Science 368:261–266. https://doi.org/10.1126/science.aat7631

Brönnimann S, Franke J, Nussbaumer SU et al (2019) Last phase of the Little Ice Age forced by volcanic eruptions. Nat Geosci 12:650–656. https://doi.org/10.1038/s41561-019-0402-y

Cavicchioli R, Ripple WJ, Timmis KN, Azam F, Bakken LR, Baylis M, Behrenfeld MJ, Boetius A, Boyd PW, Classen AT, Crowther TW, Danovaro R, Foreman CM, Huisman J, Hutchins DA, Jansson JK, Karl DM, Koskella B, Mark Welch DB, Martiny JBH, Moran MA, Orphan VJ, Reay DS, Remais JV, Rich VI, Singh BK, Stein LY, Stewart FJ, Sullivan MB, van Oppen MJH, Weaver SC, Webb EA, Webster NS (2019) Scientists' warning to humanity: microorganisms and climate change. Nat Rev Microbiol 17:569–586. https://doi.org/10.1038/s41579-019-0222-5

Ceballos G, Ehrlich PR, Barnosky AD, García A, Pringle RM, Palmer TM (2015) Accelerated modern human-induced species losses: Entering the sixth mass extinction. Sci Adv 1(5): https://doi.org/10.1126/sciadv.1400253

Ceballos G, Ehrlich PR, Raven PH (2020) Vertebrates on the brink as indicators of biological annihilation and the sixth mass extinction. Proc Natl Acad Sci U S A. 117:13596–13602. https://doi.org/10.1073/pnas.1922686117

Chaplin-Kramer R, Sharp RP, Weil C, Bennett EM et al (2019) Global modeling of nature's contributions to people. Science 366:255–258. https://doi.org/10.1126/science.aaw3372

Chen X, Tung K (2018) Global surface warming enhanced by weak Atlantic overturning circulation. Nature 559:387–391. https://doi.org/10.1038/s41586-018-0320-y

Clark MA, Domingo NGG, Colgan K, Thakrar SK, Tilman D, Lynch J, Azevedo IL, Hill JD (2020) Global food system emissions could preclude achieving the 1.5° and 2°C climate change targets. Science 370:705–708. https://doi.org/10.1126/science.aba7357

Cook J et al (2016) Consensus on consensus: a synthesis of consensus estimates on human-caused global warming. Environ Res Lett 11:4. https://doi.org/10.1088/1748-9326/11/4/048002

Cox KD, Covernton GA, Davies HL, Dower JF, Juanes F, Dudas SE (2019) Human consumption of microplastics. Environ Sci Technol 53:7068–7074

CPH 2025 Climate Plan (2011) https://stateofgreen.com/en/partners/city-of-copenhagen/solutions/copenhagen-carbon-neutral-by-2025/

De Alban JDT, Jamaludin J, de Wen DW, Than MM, Webb EL (2020) Improved estimates of mangrove cover and change reveal catastrophic deforestation in Myanmar. Environ Res Lett 15:034034

Duarte CM, Agusti S, Barbier E et al (2020) Rebuilding marine life. Nature 580:39–51. https://doi.org/10.1038/s41586-020-2146-7

EASAC (2018) Extreme weather events in Europe: preparing for climate change adaptation: an update on EASAC's 2013 study. The European Academies' Science Advisory Council

Elhacham E, Ben-Uri L, Grozovski J et al (2020) Global human-made mass exceeds all living biomass. Nature 588:442–444. https://doi.org/10.1038/s41586-020-3010-5

Fahey DW, Doherty SJ, Hibbard KA, Romanou A, Taylor PC (2017) Physical drivers of climate change. In: Wuebbles DJ, Fahey DW, Hibbard KA, Dokken DJ, Stewart BC, Maycock TK (eds) Climate science special report: fourth national climate assessment, volume I. U.S. Global Change Research Program, Washington, DC, USA, pp 73–113. https://doi.org/10.7930/j0513wcr

FAO, ITPS, GSBI, SCBD and EC (2020) State of knowledge of soil biodiversity—Status, challenges and potentialities, Report 2020. Rome, FAO. https://doi.org/10.4060/cb1928en

Fox L, Stukins S, Hill T, Miller CG (2020) Quantifying the effect of anthropogenic climate change on calcifying Plankton. Sci Rep 10:1620. https://doi.org/10.1038/s41598-020-58501-w

Ganivet E (2019) Growth in human population and consumption both need to be addressed to reach an ecologically sustainable future. Environ Dev Sustain. https://doi.org/10.1007/s10668-019-004 46-w

Gazeau F, Parker LM, Comeau S et al (2013) Impacts of ocean acidification on marine shelled molluscs. Mar Biol 160:2207–2245. https://doi.org/10.1007/s00227-013-2219-3

Gibney E (2020) Coronavirus lockdowns have changed the way Earth moves. Nature 580:176–177. https://doi.org/10.1038/d41586-020-00965-x

Gowlett JAJ (2016) The discovery of fire by humans: a long and convoluted process. Philos Trans R Soc Lond B Biol Sci 371(1696):20150164. https://doi.org/10.1098/rstb.2015.0164

Hallmann CA, Foppen RP, van Turnhout CA, de Kroon H, Jongejans E (2014) Declines in insectivorous birds are associated with high neonicotinoid concentrations. Nature 511(7509):341–343. https://doi.org/10.1038/nature13531

Hansen JE, Sato M (2012) Paleoclimate implications for human-made climate change. In: Berger A, Mesinger F, Sijacki D (eds) Climate change. Springer, Vienna, pp 21–47

Hoegh-Guldberg O, Jacob D, Taylor M, Guillén Bolaños T, Bindi M, Brown S, Camilloni IA, Diedhiou A, Djalante R, Ebi K, Engelbrecht F, Guiot J, Hijioka Y, Mehrotra S, Hope CW, Payne AJ, Pörtner HO, Seneviratne SI, Thomas A, Warren R, Zhou G (2019) The human imperative of stabilizing global climate change at 1.5°C. Science 365:eaaw6974. https://doi.org/10.1126/science.aaw6974

Hönisch B, Ridgwell A, Schmidt DN, Thomas E, Gibbs SJ, Sluijs A, Zeebe R, Kump L, Martindale RC, Greene SE, Kiessling W, Ries J, Zachos JC, Royer DL, Barker S, Marchitto TM Jr, Moyer R, Pelejero C, Ziveri P, Foster GL, Williams B (2012) The geological record of ocean acidification. Science 335:1058–1063. https://doi.org/10.1126/science.1208277

Hulme M (2009) On the origin of 'the greenhouse effect': John Tyndall's 1859 interrogation of nature. Weather 64:121–123. https://doi.org/10.1002/wea.386

Institute for Economics & Peace. Ecological Threat Register 2020: Understanding Ecological Threats, Resilience and Peace. Sydney, September 2020. Available from: http://visionofhuma nity.org/reports (accessed 08 June 2021)

IPCC (2014) Climate change 2013: the physical science basis Working Group I contribution to the fifth assessment report of the intergovernmental panel on climate change. WMO/UNEP

IPCC (2018) IPCC, Global warming of 1.5 °C—an IPCC special report on the impacts of global warming of 1.5 °C above pre-industrial levels and related global greenhouse gas emission pathways, in the context of strengthening the global response to the threat of climate change. https://www.ipcc.ch/sr15/

IPCC (2019) IPCC Special report on the ocean and cryosphere in a changing climate. In Pörtner H-O, Roberts DC, Masson-Delmotte V, Zhai P, Tignor M, Poloczanska E, Mintenbeck K, Alegría A, Nicolai M, Okem A, Petzold J, Rama B, Weyer NM (eds). Ipcc 2019 https://www.ipcc.ch/srocc/

IPBES (2019) Global assessment report on biodiversity and ecosystem services of the Intergovernmental Science-Policy Platform on Biodiversity and Ecosystem Services. Brondizio ES, Settele J, Díaz S, Ngo HT (eds). IPBES secretariat, Bonn, Germany

Jiang L, Carter BR, Feely RA et al (2019) Surface ocean pH and buffer capacity: past, present and future. Sci Rep 9:18624. https://doi.org/10.1038/s41598-019-55039-4

Keeling CD, Mook WG, Tans PP (1979) Recent trends in the 13C/12C ratio of atmospheric carbon dioxide. Nature 277:121–123

Kelly LT, Giljohann KM, Duane A, Aquilué N, Archibald S, Batllori E, Bennett AF, Buckland ST, Canelles Q, Clarke MF, Fortin MJ, Hermoso V, Herrando S, Keane RE, Lake FK, McCarthy MA, Morán-Ordóñez A, Parr CL, Pausas JG, Penman TD, Regos A, Rumpff L, Santos JL, Smith AL, Syphard AD, Tingley MW, Brotons L (2020) Fire and biodiversity in the Anthropocene. Science 370(6519):eabb0355. https://doi.org/10.1126/science.abb0355

Le Quéré C, Jackson RB, Jones MW, Smith Adam J P, Abernethy S, Andrew RM, De-Gol AJ, Willis DR, Shan Y, Canadell JG, Friedlingstein P, Creutzig F, Peters GP (2020) Temporary reduction in daily global CO$_2$ emissions during the COVID-19 forced confinement. Nature Climate Change 10:647–653. https://doi.org/10.1038/s41558-020-0797-x

Lecocq T, Hicks SP, Van Noten K, van Wijk K, Koelemeijer P, De Plaen RSM, Massin F, Hillers G, Anthony RE, Apoloner MT, Arroyo-Solórzano M, Assink JD, Büyükakpınar P, Cannata A, Cannavo F, Carrasco S, Caudron C, Chaves EJ, Cornwell DG, Craig D, den Ouden OFC, Diaz J, Donner S, Evangelidis CP, Evers L, Fauville B, Fernandez GA, Giannopoulos D, Gibbons SJ, Girona T, Grecu B, Grunberg M, Hetényi G, Horleston A, Inza A, Irving JCE, Jamalreyhani M, Kafka A, Koymans MR, Labedz CR, Larose E, Lindsey NJ, McKinnon M, Megies T, Miller MS, Minarik W, Moresi L, Márquez-Ramírez VH, Möllhoff M, Nesbitt IM, Niyogi S, Ojeda J, Oth A, Proud S, Pulli J, Retailleau L, Rintamäki AE, Satriano C, Savage MK, Shani-Kadmiel S, Sleeman R, Sokos E, Stammler K, Stott AE, Subedi S, Sørensen MB, Taira T, Tapia M, Turhan F, van der Pluijm B, Vanstone M, Vergne J, Vuorinen TAT, Warren T, Wassermann J, Xiao H (2020) Global quieting of high-frequency seismic noise due to COVID-19 pandemic lockdown measures. Science 369:1338–1343. https://doi.org/10.1126/science.abd2438

Lenton TM, Rockström J, Gaffney O, Rahmstorf S, Richardson K, Steffen W, Schellnhuber HJ (2019) Climate tipping points—too risky to bet against. Nature 575:592–595. https://doi.org/10.1038/d41586-019-03595-0

Levitus S, Antonov J, Boyer T, Baranova O, Garcia H, Locarnini R, Mishonov A, Reagan J, Seidov D, Yarosh E, Zweng M (2017) NCEI ocean heat content, temperature anomalies, salinity anomalies, thermosteric sea level anomalies, halosteric sea level anomalies, and total steric sea level anomalies from 1955 to present calculated from in situ oceanographic subsurface profile data (NCEI Accession 0164586). Version 4.4. NOAA National Centers for Environmental Information. Dataset. https://doi.org/10.7289/v53f4mvp

Liu Z, Ciais P, Deng Z et al (2020) Near-real-time monitoring of global CO2 emissions reveals the effects of the COVID-19 pandemic. Nat Commun 11:5172. https://doi.org/10.1038/s41467-020-18922-7

Lüthi D, Le Floch M, Bereiter B, Blunier T, Barnola J-M, Siegenthaler U, Raynaud D, Jouzel J, Fischer H, Kawamura K, Stocker TF (2008) High-resolution carbon dioxide concentration record 650,000-800,000 years before present. Nature 453:379–382. https://doi.org/10.1038/nature06949

Macreadie PI, Anton A, Raven JA et al (2019) The future of Blue Carbon science. Nat Commun 10:3998. https://doi.org/10.1038/s41467-019-11693-w

Milner AM, Khamis K, Battin TJ, Brittain JE, Barrand NE, Füreder L, Cauvy-Fraunié S, Gíslason GM, Jacobsen D, Hannah DM, Hodson AJ, Hood E, Lencioni V, Ólafsson JS, Robinson CT (2017) Glacier shrinkage driving global changes in downstream systems. Proc Natl Acad Sci 114(37):9770–9778

Mikaloff-Fletcher SE, Schaefer H (2019) Rising methane: a new climate challenge. Science 364:932–933

Myllyvirta L, Theriot H (2020) 11,000 air pollution-related deaths avoided in Europe as coal, oil consumption plummet. Center for Research on Energy and Clean Air Report. https://energyand cleanair.org/wp/wp-content/uploads/2020/04/CREA-Europe-COVID-impacts.pdf

National Academy of Sciences (2020) Climate change: evidence and causes: Update 2020. Washington, DC: The National Academies Press. https://doi.org/10.17226/25733

Nerem RS, Beckley BD, Fasullo JT, Hamlington BD, Masters D, Mitchum GT (2018) Climate-change–driven accelerated sea-level rise detected in the altimeter era. PNAS. https://doi.org/10.1073/pnas.1717312115

Neukom R, Steiger N, Gómez-Navarro JJ, Wang J, Werner JP (2019a) No evidence for globally coherent warm and cold periods over the preindustrial common era. Nature 571:550–554. https://doi.org/10.1038/s41586-019-1401-2

Neukom R, Barboza LA, Erb MP et al (2019b) Consistent multidecadal variability in global temperature reconstructions and simulations over the Common Era. Nat Geosci 12:643–649. https://doi.org/10.1038/s41561-019-0400-0

Pan Y, Birdsey RA, Phillips OL, Jackson RB (2013) The structure, distribution, and biomass of the world's forests. Annu Rev Ecol Evol Syst 44:593–622

Potts SG, Neumann P, Vaissière B, Vereecken NJ (2018) Robotic bees for crop pollination: why drones cannot replace biodiversity. Sci Total Environ 642:665–667. https://doi.org/10.1016/j.scitotenv.2018.06.114

Renssen R, Seppä H, Heiri O, Roche DM, Goosse H, Fichefet T (2009) The spatial and temporal complexity of the Holocene thermal maximum. Nat Geosci 2:411–414

Román-Palacios C, Wiens JJ (2020) Recent responses to climate change reveal the drivers of species extinction and survival. Proc Natl Acad Sci USA 117:4211–4217. https://doi.org/10.1073/pnas.1913007117

Rosenberg KV, Dokter AM, Blancher PJ, Sauer JR, Smith AC, Smith PA, Stanton JC, Panjabi A, Helft L, Parr M, Marra PP (2019) Decline of the North American Avifauna. Science 366:120–124. https://doi.org/10.1126/science.aaw1313

RSPB (2020) A lost decade for nature. http://ww2.rspb.org.uk/Images/A%20LOST%20DECADE%20FOR%20NATURE_tcm9-481563.pdf

Sasgen I, Wouters B, Gardner AS et al (2020) Return to rapid ice loss in Greenland and record loss in 2019 detected by the GRACE-FO satellites. Commun Earth Environ 1:8. https://doi.org/10.1038/s43247-020-0010-1

Service RF (2020) The carbon vault. Science 369:1156–1159

Schelske O, Wilke B, Retsa A, Rutherford-Liske G, de Jong R (2020) Biodiversity and ecosystems services: a business case for re/insurance. Swiss Re Institute Expertise Publication

Shakun JD (2018) Pollen weighs in on a climate conundrum. Nature 554:39–40. https://doi.org/10.1038/d41586-018-00943-4

Shannon G, McKenna MF, Angeloni LM, Crooks KR, Fristrup KM, Brown E, Warner KA, Nelson MD, White C, Briggs J, McFarland S, Wittemyer G 2016. A synthesis of two decades of research documenting the effects of noise on wildlife. Biol Rev Camb Philos Soc 91, 982–1005. https://doi.org/10.1111/brv.12207

Sherwood SC, Huber M (2010) An adaptability limit to climate change due to heat stress. PNAS 107:9552–9555. https://doi.org/10.1073/pnas.0913352107

Smith B, Fricker HA, Gardner AS, Medley B, Nilsson J, Paolo FS, Holschuh N, Adusumilli S, Brunt K, Csatho B, Harbeck K, Markus T, Neumann T, Siegfried MR, Zwally HJ (2020) Pervasive ice sheet mass loss reflects competing ocean and atmosphere processes. Science eaaz5845. https://doi.org/10.1126/science.aaz5845

Soroye P, Newbold T, Kerr J (2020) Climate change contributes to widespread declines among bumble bees across continents. Science 367:685–688

Spalding MD, Brown BE (2015) Warm-water coral reefs and climate change. Science 350:769–771

Stillman JH (2019) Heat waves, the new normal: summertime temperature extremes will impact animals, ecosystems, and human communities. Physiology (Bethesda) 34:86–100. https://doi.org/10.1152/physiol.00040.2018

Storlazzi CD, Gingerich SB, van Dongeren A, Cheriton OM, Swarzenski PW, Quataert E, Voss CI, Field DW, Annamalai H, Piniak GA, McCall R (2018) Most atolls will be uninhabitable by the mid-21st century because of sea-level rise exacerbating wave-driven flooding. Sci Adv 4(4):eaap9741. https://doi.org/10.1126/sciadv.aap9741

Subramanian M (2019) Anthropocene now: influential panel votes to recognize Earth's new epoch. Nature 21 May 2019. https://doi.org/10.1038/d41586-019-01641-5

Tape KD, Christie K, Carroll G, O'Donnell JA (2016) Novel wildlife in the Arctic: the influence of changing riparian ecosystems and shrub habitat expansion on snowshoe hares. Glob Chang Biol 22(1):208–219. https://doi.org/10.1111/gcb.13058

Thomas N, Nigam S (2018) Twentieth-century climate change over Africa: seasonal hydroclimate trends and Sahara desert expansion. J Clim 31:3349–3370

Tranter M, Brown LE (2017) Glacier shrinkage driving global changes in downstream systems. PNAS USA 114:9770–9778. https://doi.org/10.1073/pnas.1619807114

UN Environmental Program Foresight Brief (2019) https://wedocs.unep.org/bitstream/handle/20.500.11822/28453/Foresight013.pdf

UN Global Diversity Report 5 (2020) https://www.cbd.int/gbo/gbo5/publication/gbo-5-en.pdf

Velicogna I, Mohajerani Y, Geruo A, Landerer F, Mouginot J, Noel B, Rignot E, Sutterley T, van den Broeke M, van Wessem M, Wiese D (2020) Continuity of ice sheet mass loss in Greenland and Antarctica from the GRACE and GRACE Follow-on missions. Geophys Res Lett 47:e2020GL087291. https://doi.org/10.1029/2020gl08729

Vicedo-Cabrera AM, Guo Y, Sera F et al (2018) Temperature-related mortality impacts under and beyond Paris Agreement climate change scenarios. Climatic Change 150:391–402. https://doi.org/10.1007/s10584-018-2274-3

Viglione G (2020) Climate lawsuits are breaking new legal ground to protect the planet. Nature 579:184–185. https://doi.org/10.1038/d41586-020-00175-5

Watson JEM, Shanahan DF, Di Marco M, Allan J, Laurance WF, Sanderson EW, Mackey B, Venter O (2016) Catastrophic declines in wilderness areas undermine global environment targets. Curr Biol 26:2929–2934. https://doi.org/10.1016/j.cub.2016.08.049

Williams BA, Venter O, Allan JR, Atkinson SC, Rehbein JA, Ward M, Di Marco M, Grantham HS, Ervin J, Goetz SJ, Hansen AJ, Jantz P, Pillay R, Rodríguez-Buriticá S, Supples C, Virnig ALS, Watson JEM (2020) Change in terrestrial human footprint drives continued loss of intact ecosystems. One Earth 3:371–382. https://doi.org/10.1016/j.oneear.2020.08.009

WWF (2018) Living planet report—2018: Aiming Higher. In: Grooten M, Almond REA (eds). WWF, Gland, Switzerland

WWF (2020) Living planet report 2020—bending the curve of biodiversity loss. In: Almond REA, Grooten M, Petersen T (eds). WWF, Gland, Switzerland

Xu C, Kohler TA, Lenton TM, Svenning JC, Scheffer M (2020) Future of the human climate niche. Proc Natl Acad Sci U S A. 117:11350–11355. https://doi.org/10.1073/pnas.1910114117

Xu J, Grumbine RE, Shrestha A, Eriksson M, Yang X, Wang Y, Wilkes A (2009) The melting Himalayas: cascading effects of climate change on water, biodiversity, and livelihoods. Conserv Biol 23:520–530. https://doi.org/10.1111/j.1523-1739.2009.01237.x

Chapter 9
Artificial Intelligence: Opportunity or Risk?

Artificial intelligence (AI) is a very hot topic. This field considers that new computing technologies can be developed to a level of intelligence far beyond that of the human brain. Many books have been written on AI, some of which are bestsellers: for example, those by Max Tegmark (2017), Nick Bostrom (2014), and Ranga Yogeshwar (2017). In addition, there are frequent discussions and predictions about the future of AI in the news media. Most striking to JE was a lecture by the philosopher Thomas Metzinger (2018). He predicted that AI entities could potentially make a decision to eliminate humans. This is a logical scenario in view of the many problems that humans have caused on Earth, as described in Chaps. 7 and 8. Of course, these predictions are, at present, extremely speculative and, as we will discuss below, there are many potential useful applications of AI. However, workers in the field of AI are convinced that these new powers of computers will bring massive, unprecedented changes and dangers as well as benefits to human life. Indeed, Max Tegmark stated: "If future superhuman artificial intelligence becomes the biggest event in human history, how then can we ensure that it doesn't become the last?" (2017).

9.1 What Is Intelligence and What Is AI?

The current period of human history has often been termed the digital age because of the domination of digital technologies. This book is written on a computer. The computer's connection to data files in the internet helps to find information and keep the references to the scientific literature up to date. It is a popular perception that the younger generation are unable to live without their cell phones in hand, and this is often viewed as a limitation. However, the growth of mobile phone use in Africa, for example, has brought tangible benefits of vastly improved telecommunications without the need to build new landline infrastructure. Shepherds in Africa and farmers in India now use cell phones to organize their work and finances. All of this computer technology has been developed from human curiosity and problem

© The Author(s), under exclusive license to Springer Nature Switzerland AG 2021
J. C. Adams and J. Engel, *Life and Its Future*,
https://doi.org/10.1007/978-3-030-59075-8_9

solving. Charles Babbage (1791–1871) in the UK is credited with designing the first automated machine to solve mathematical calculations, his so-called "Analytical Engine", but this machine was never built (https://www.bbc.co.uk/history/historic_figures/babbage_charles.shtml).

All the abilities of a standard modern computer depend on routines and algorithms which are designed and programmed by humans. For example, when we ask our computer for the volume V of a sphere with radius r of 1 cm it will look up the programmed relation $V = 4/3 \, \pi \, r^3$, find the value of $\pi = 3.14159$ in a programmed file of natural numbers, insert the value of r and come up with the result $V = 4.18879 \, cm^3$ by application of the (also programmed) multiplication rules. For more complicated tasks, the programmed routines, called software or algorithms in computer language, are more complex. For example, to play chess, a conventional computer programme must include the rules of the game and also strategies of attack and defence that have been applied successfully by human chess experts, along with evaluations of how the game might continue after a particular move. In both of these examples, the intelligence comes from the human programmer and is executed by the computer.

In 1997, the Deep Blue computer built by IBM was the first computer to beat the world chess champion, Garry Kasparov, in a six-match game. Thus, computers built as described above can be very efficient. In addition to efficient software, computers generally offer a high speed of calculation and a high reliability of execution. These properties are essential for the solution of complicated problems. Another example of the utility of computers is the solution of differential equations for which no algebraic solution exists, by iteration processes. The iteration processes are well-known mathematical methods, but their manual execution can be extremely slow. Nowadays, fast computers, for example Math10 Banners (2019), can solve complicated mathematical problems within seconds. These computers still provide answers based on pre-defined rules.

A completely novel feature became added to the field when computers were developed that were no longer dependent on pre-designed software but were able to develop their own software (from data provided) which was better than that provided by humans. This developmental ability is called artificial intelligence (AI) or machine-learning. A much-discussed example is the AI computer AlphaGo Zero, which mastered the board game of Go without prior knowledge provided by humans (Silver et al. 2017) (Fig 9.1). Very impressive computer methods were developed to set up the internally developed programs and learning by the computer. These methods are summarized under terms such as "deep learning" or "neural networks". In addition to game playing, deep learning has been advanced to "mine" the increasingly huge amounts of data that are available in the world. In the biosciences, around 10,000 research papers are published online every day, and many of these contain large datasets relating to images, gene expression profiles, proteomics, or high-throughput screening of candidate drugs. No one person alone can assimilate all these data, but deep-learning algorithms can compare and statistically analyse datasets to extract patterns or correlations without pre-conceptions. These can fuel future hypotheses and laboratory experiments by researchers.

Fig. 9.1 A board of the Go game. White plays against Black. A player can place their stone on any intersection of the network. The winner is the player who ends up with the maximum territory surrounded by stones

AlphaGo Zero played against a number of top human players in the Go game and was the winner in most cases. From this observation many enthusiasts derived the conclusion that AI offers higher levels of intelligence than human intelligence. In discussions with developers of AI, JE learned that, even for a computer expert, it is often difficult to understand the novel computing routines developed by the computer. This raises an inherent issue: if we cannot understand how the solutions were generated, how can we trust and interpret the results? Computer scientists refer to the concept of a point at which an artificial intelligence could go beyond human control as the "singularity". The singularity point is thought to be at least several decades away; nevertheless, overall, there are already major ethical questions and issues around the implementation of AI and our use of its outputs.

AI is a new field and many of the current conclusions may be modified as experience grows. At the present time there are many arguments against an overestimation of AI. The cost of one AlphaGo Zero system is about 25 million dollars (Wikipedia, 2019) and the first version could only play Go. In this sense, the human brain remains much more versatile. However, this is a rapidly moving frontier. More recently, the property of AlphaGo Zero to learn by reinforcement was adapted into a more general algorithm, in which the computer "learns" by playing against itself once given the basic rules of a new game. With this self-learning activity, AlphaZero was able to teach itself chess to world champion levels in a matter of hours, thus showing a super-human learning capacity (Silver et al. 2018).

The potential of AI is currently under intense discussion by philosophers, scientists, computer experts, technology developers and politicians as well as the general public. In relation to the biosciences, examples of areas where AI is seen as bringing potential benefits (of cost, timeliness and accuracy) would be the ability to predict a protein's structure from its sequence of amino acids, to identify a cell signalling pathway that drives a particular disease process, or the most suitable category of small molecule for a new therapeutic strategy (The Biochemist Magazine, 2019). The prediction of protein structures from amino acid sequence has been a "holy grail" of biochemists and biophysicists over the last fifty years, given the expense and length of time that is needed to solve a structure experimentally from protein crystals. International trials termed CASP (critical assessment of structure prediction; predictioncenter.org) encourage research groups to test their modelling softwares in an unbiased way against a set of sequences of unknown structure. CASP trials have been held regularly since 1994 to assess how the field is progressing and which modelling approaches are the most promising. At CASP13 in 2018, a deep-learning method named AlphaFold, that is based on training of neural networks, proved the most successful method, by generating accurate structures for 24 out of 43 protein domains in the test (Senior et al. 2020). AlphaFold was developed by the Deep-Mind company which was set up in London in 2010 (deepmind.com). At CASP14 in 2020, the latest version of AlphaFold, trained on all known protein structures, delivered a stunning breakthrough in predicting unknown protein structures with an accuracy comparable to experimentally-determined structures (Jumper et al. 2020). Structural biologists have hailed this as a game-changing breakthrough for the entire field (Service, 2020).

The ability of AI to absorb massive amounts of data to generate novel solutions (that might have eluded human researchers), to problems, was also made dramatically clear from a recent application of AI to identify new candidate antibiotic molecules. As mentioned in Chap. 6, antibiotic-resistant bacteria have become a major threat to human health, especially in hospitals. The antibiotics in use today were identified many years ago and it has proved difficult for conventional research to identify new types of antibiotics against which bacteria have no resistance. In the study, the first of its kind, a deep neural network was trained to predict molecules with anti-bacterial activity and then given the task of screening 6,000 compounds that are already under investigation for activity against human diseases. The AI network searched the library for compounds using its own prediction of which might have anti-bacterial activity, and also looked for molecules with different chemical structures to current antibiotics. This process identified an entirely new antibiotic, halicin, which researchers then found to have strong killing activity against a range of bacteria under laboratory conditions. Next, the researchers set the deep neural network to screen a library of 107 million compound structures. Twenty-three candidate compounds were identified within 3 days, two of which have promising anti-bacterial activity in laboratory tests (Stokes et al. 2020). This scale and speed of screening is far beyond that of conventional computers.

Overall, AI has great potential for a huge scope of clinical applications. Through deep-learning methods, computers could scan millions of images and "learn" from the features of the samples to identify diseased tissue in comparison to the normal tissue. For example, at present, individuals with diabetes who are at risk of retinal degeneration (a pathology termed diabetic retinopathy) have images taken of their retinas. These images are assessed manually by experienced ophthalmologists which is a time-consuming process. If AI could be used to screen the images to distinguish people with moderate to severe diabetic retinopathy for referral to a human doctor, monitoring could be carried out more quickly and cheaply, and the ophthalmologists' time could be focused onto the patients with the most urgent needs. Clearly, there are ethical concerns around delegating the first stages of diagnosis to a machine and many independent tests would be needed to assess the accuracy of the computer-based scoring of images. However, a trial of a deep-learning algorithm produced encouraging results with an accuracy of detecting retinopathy comparable to that of medical experts (De Fauw et al. 2018). Furthermore, analysis by computers rather than humans may have benefits of consistency. An algorithm will always interpret the same sample in the same way, and will not suffer from fatigue, as a human doctor may. Clinicians also have strong interests in AI technology to improve the accuracy of cancer detection in biopsy samples (ie, to minimise false positive or false negative results in detecting a tumour) and the classification of sub-types of tumours from a particular tissue. For example, by combining deep learning from images with pre-training of computers on libraries of images with annotated tumours, it might be possible to achieve more rapid, accurate detection of breast cancer from mammograms (Shen et al. 2019).

While these uses of deep-learning are seen as beneficial, more controversial is the collection and analysis of large-scale data on people's movements or purchasing choices conducted by organisations such as Google or Amazon. Incorporation of AI into facial recognition technology or behaviour prediction can provide governments with unprecedented scope for national surveillance. These activities and their future possibilities have opened up major new questions about personal privacy and the legal frameworks in which the providers of internet platforms operate.

AI is also driving new directions for the development of mechanical robots that can handle sophisticated tasks. Very dangerous risks could be associated with military applications of AI, such as the concept of "robot soldiers" with capacity for automatic facial recognition. How controllable would these be when released and what parameters would be deemed satisfactory for target recognition? Other applications of AI in robotics appear to have beneficial potential. The concept of robot "pets" as calming companions for elderly people who are not well enough to look after a living pet, or who suffer from allergies or dementia, is gaining ground (Bates 2019). Another development would be AI talking robots designed as "social assistants" to hold conversations, prompt elderly people to take medicines, keep physically active and play games, whilst acting as "helpers" to carry out domestic chores (Matuszek 2017). This area of robotics is of particular interest in countries where a shortfall of carers for the increasing number of elderly people is predicted over the next years.

The development of smart robots also relates to an area termed trans-humanism. The proponents of trans-humanism propose to remove human ageing as a cause of death, by expanding on technology for replacement organs or limbs and also by achieving enhanced brain capacity. For example, cognitive decline in later life could be halted or abolished by implanting memory chips or AI capacity into a person's body. The full vision of trans-humanism is to produce "super-humans", in which the biological body is merged with machines and AI capacity. These beings would be stronger and more robust than humans, with higher intelligence and extended lifespans (Kurzweil 2000). These semi-inorganic, potentially eternal beings would have increased capacity to take journeys into space, thus making it more feasible for humans to leave Earth and colonise other planets. At present, the technology to build a trans-human does not exist and obviously the envisioned scope of the modifications to the human body raises many new ethical questions. Some enthusiasts have promoted the idea of having their bodies deep-frozen in liquid nitrogen ($-196\ ^{\circ}$C) immediately after legal death, so that they can be preserved until a time that suitable technology is available for them to be "converted" to trans-humans (O'Connell 2017). The largest cryonics facility is that of the Alcor Life Extension Foundation in Arizona (https://www.alcor.org).

No day passes without an AI lecture on television or radio or a publication in a journal. Among the different authors, thinkers such as the astrophysicist Sir Stephen Hawking (2018), have warned that an unwanted take-over by AI (i.e., by origination of AI machines that could go beyond human control) would be disastrous. It also follows from the textbooks of Bostrom (2014) and Tegmark (2017) that, at present, negative expectations outweigh positive achievements from AI. Many visions describe a situation in which humans become helpless in the face of a powerful AI system that cannot be switched off, or that gains the capacity to cause catastrophic changes in the world very rapidly. An extreme example is the already mentioned vision of Thomas Metzinger (2018), according to which human beings could be exterminated after a take-over by AI entities.

Discussions about the pros and cons of AI often mix this new computer technology with other advanced computer techniques. Self-driving cars, robots, automatic facial recognition and connection of digital systems to the brain and nervous system are all hot topics. In the healthcare market, Google Deepmind Health is focusing on applying AI models to advanced healthcare. The company believes that their technology will be able to develop entirely new diagnostic procedures in the near future. A review (Ijaz 2018) noted ten companies that are working in similar directions. Several medical conventions have taken place with the aim to revolutionize medicine through AI. The scientific publisher Elsevier began the journal "Artificial Intelligence in Medicine" in 1989 and there are now many scientific journals related to AI and healthcare. Commercial interest in new inventions is frequently very high, but AI beats all records. The scientific foundations of this enthusiasm are still very limited and many problems and risks are connected with applications of AI.

9.2 Legal Problems

As a scenario, suppose that a robot soldier kills a person because of an error in facial recognition. Who is responsible and who is guilty? Our legal systems assumes that a crime is committed by sentient persons and not by machines or AI computers. A similar problem will arise when a self-driving car is involved in a traffic accident.

The need for a completely novel legal system to deal with AI and its implications is discussed in depth in the books of Tegmark (Tegmark (2017), Bostrom (2014) and Metzinger (2018). To address this issue in the UK, the UK Information Commissioner's Office has started to develop AI auditing framework guidance for organisations that have begun to use AI. The goal is to ensure that these organisations must explain clearly how their technology functions to anyone who will be affected by its application (https://ico.org.uk/media/about-the-ico/consultations/2617219/guidance-on-the-ai-auditing-framework-draft-for-consultation.pdf).

9.3 What Is Progress?

In the afterword of his book, Nick Bostrom (2014) complained about the strength of public criticisms of AI. He argued that this could result in failures of research funding for this field and could prevent progress. This touches on a very general problem. The inventors of a new technique are normally curious and interested to push their invention to final success. It is a common experience that scientific or engineering progress is connected with many problems and risks. Chapter 4 describes many examples and AI is certainly no exception. Researchers and inventors need to consider what predictable risks might arise from their discoveries. Many AI experts have done so, and we have cited examples of their very critical predictions. As with any new technology (for example, the beginnings of molecular biology), AI will be subjected to ethical scrutiny, governmental regulation and legal frameworks.

A very strong driving force for new inventions is the possibility of commercial applications. Advertisement and enticement are highly developed business areas. Together with the big data analysis of human behaviour, this is leading to new types of aggressive marketing. Currently artificial intelligence is a top key word in marketing. In fact, often the use of the term AI is not accurate, and the term "pseudo AI" is sometimes used. We have seen the implication of AI in advertisements of novel health care, healthy food products, beauty products, routes to happiness, computer games, family care, cell phones and education, and this is not an exhaustive list.

AI also poses questions and challenges to our understanding of artistic creativity. Artists, composers and writers stand out in human society because of their unique creative articulation of deep insights, feelings and emotions that speak to the human spirit and experience. Yet now, for example, AI music software can generate perfect imitations of classical and modern composers. DeepBach (Vincent 2016) is a published example. From such imitations, it is only another step to AI that could generate original music that is as memorable and beautiful as that of the most famous composers. Could we view such products as creative as well as technological achievements? If we know that a poem or a symphony was written by a computer, will we have the same emotional response to it as to a poem or music written by a human? Artistic endeavours are a human activity of high value to individuals and societies. We should take great care not to lose appreciation of this capacity. In counterbalance, there is also a view that AI may offer "augmented creativity": that is, because of the unbiased approach of AI in deep learning, novel creative possibilities and ideas may emerge that can then be taken up by creative artists.

It is clear that there are many risks, as well as potential benefits, connected with deep learning and AI, of which the loss of each person's own intellectual independence and privacy could be a very severe effect. How might further technical progress in AI address, instead of adding to, the general problems of our world? Jonathan Ledgard is a journalist, novelist and technologist who has begun to consider AI in relation to nature and the ongoing loss of biodiversity (as discussed in Chap. 8). He has suggested that including awareness of the complexity of all living species and the value of life in the databanks of AI machines could be a route to finding new solutions to address habitat loss and other ecological issues (Taub 2019). Another ongoing initiative is that of the OpenAI company in San Francisco that has the goal to develop "artificial general intelligence" (AGI) that is safe to use and can benefit all people; for example, by applying AGI to very large-scale problems such as climate change (https://openai.com/charter/).

References

Bates M (2019) Robotic pets: a senior's best friend? https://pulse.embs.org/july-2019/robotic-pets-a-seniors-best-friend/

Bostrom N (2014) Superintelligence, paths, dangers, strategies. Oxford University Press

De Fauw J, Ledsam JR, Romera-Paredes B, Nikolov S, Tomasev N, Blackwell S, Askham H, Glorot X, O'Donoghue B, Visentin D, van den Driessche G, Lakshminarayanan B, Meyer C, Mackinder F, Bouton S, Ayoub K, Chopra R, King D, Karthikesalingam A, Hughes CO, Raine R, Hughes J, Sim DA, Egan C, Tufail A, Montgomery H, Hassabis D, Rees G, Back T, Khaw PT, Suleyman M, Cornebise J, Keane PA, Ronneberger O (2018) Clinically applicable deep learning for diagnosis and referral in retinal disease. Nat Med 24:1342–1350

Hawking S (2018) Brief answers to the big questions: the final book from Stephen Hawking. John Murray

Ijaz R (2018) https://bigdata-madesimple.com/10-most-innovative-ai-companies

Jumper J, Evans R, Pritzel A, Green T, Figurnov M, Tunyasuvunakool K, Ronneberger O, Bates R, Žídek A, Bridgland A, Meyer C, Kohl SAA, Potapenko A, Ballard AJ, Cowie A, Romera-Paredes B, Nikolov S, Jain R, Adler J, Back T, Petersen S, Reiman D, Steinegger M, Pacholska M, Silver D, Vinyals O, Senior AW, Kavukcuoglu K, Kohli P, Hassabis D (2020) High accuracy protein structure prediction using deep learning. In: Fourteenth critical assessment of techniques for protein structure prediction (Abstract Book), 30 November–4 December 2020

Kurzweil R (2000) The age of spiritual machines: when computers exceed human intelligence. Penguin Books

Math10 Banners (2019) https://www.math10.com/link-to-us.html

Matuszek C (2017) How robots could help the elderly age in their homes. Smithsonianmag.com/innovation, August 29, 2017

Metzinger T (2018) Virtuelle Realität und Künstliche Intelligenz. Collegium Helveticum

O'Connell M (2017) To be a machine: adventures among cyborgs, utopians, hackers, and the futurists solving the modest problem of death. Doubleday Books

Senior AW, Evans R, Jumper J, Kirkpatrick J, Sifre L, Green T, Qin C, Žídek A, Nelson AWR, Bridgland A, Penedones H, Petersen S, Simonyan K, Crossan S, Kohli P, Jones DT, Silver D, Kavukcuoglu K, Hassabis D (2020) Improved protein structure prediction using potentials from deep learning. Nature 577(7792):706–710. https://doi.org/10.1038/s41586-019-1923-7

Service (2020) The game has changed: AI triumphs at protein folding. Science 370:1144–1145

Shen L, Margolies LR, Rothstein JH, Fluder E, McBride R, Sieh W (2019) Deep learning to improve breast cancer detection on screening mammography. Sci Rep 9(Article number: 12495)

Silver D, Schrittwieser J, Simonyan K, Antonoglou I, Huang A, Guez A, Hubert T, Baker L, Lai M, Bolton A, Chen Y, Lillicrap T, Hui F, Sifre L, Van Den Driessche G, Graepel T, Hassabis D (2017) Mastering the game of Go without human knowledge. Nature 550:354–359

Silver D, Hubert T, Schrittwieser J, Antonoglou I, Lai M et al (2018) A general reinforcement learning algorithm that masters chess, shogi, and Go through self-play. Science 362:1140–1144. https://doi.org/10.1126/science.aar6404

Stokes JM, Yang K, Swanson K, Jin W, Cubillos-Ruiz A, Donghia NM, MacNair CR, French S, Carfrae LA, Bloom-Ackerman Z, Tran VM, Chiappino-Pepe A, Badran AH, Andrews IW, Chory EJ, Church GM, Brown ED, Jaakkola TS, Barzilay R, Collins JJ (2020) A deep learning approach to antibiotic discovery. Cell 180:688-702.e13

Taub, B (2019). Jonathan Ledgard believes imagination could save the world. September 23, 2019 issue, New Yorker Magazine

Tegmark M (2017) Life 3.0, being human in the age of Artificial Intelligence. Penguin Books

Vincent J (2016) Can you tell the difference between Bach and Robobach? The Verge https://www.theverge.com/2016/12/23/14069382/ai-music-creativity-bach-deepbach-csl

Yogeshwar R (2017) Nächste Ausfahrt Zukunft, Oertli

Chapter 10
Life on Other Planets

10.1 Could There Be Life-Forms on Other Planets?

The possibility that there could be life on other planets is an old and much discussed topic of astronomy that previously went under the name of the search for extraterrestrial intelligence (SETI). Up to now there is no direct proof for the existence of life outside our own planet (Enrico Fermi remarked that if another intelligent life form was out there we would have met it by now), but astronomers and others have defined the conditions under which life may exist. A suitable temperature range and physiochemical conditions would be required on the planet, along with a solvent such as water, an energy source and a suitable range of elements (Cockell et al. 2016).

A related question is whether life on other planets would also be carbon-based (see Chap. 3), or could instead be based on other elements. The idea of a different chemistry of life based on silicon has been discussed over many years. The element silicon is more common on Earth than carbon and, like carbon, can form chemical bonds with up to four other atoms, giving potential for building complex molecules. Various organisms on Earth, including diatoms, certain sponges and some plants have evolved enzymes to form silicon-dioxide-based biomineralised structures that support tissues or provide protective exoskeletons (Schröder et al. 2008). However, overall, silicon chemistry results in less complex molecules than carbon. In addition, silicon has a great tendency to bond with oxygen, which means that, on Earth, most silicon is bound up in rocks (Jacob 2016). There are no known natural examples of molecules that contain both silicon and carbon (referred to as organosilicon), but chemists have developed synthetic methods for various organosilicons that have many useful industrial applications.

Recent research on organosilicon formation in the laboratory has given insights into the reaction conditions that are needed. Researchers identified that haemin (a protoporphyrin compound containing a Fe^{3+} ion) and several natural cytochrome c proteins (cytochrome c is a haem-containing protein important for electron transfer in mitochondria) have a fortuitous capacity to drive organosilicon formation by carbene ($R = C$) insertion into silicon-hydrogen bonds under physiological conditions in

J. C. Adams and J. Engel, *Life and Its Future*,
https://doi.org/10.1007/978-3-030-59075-8_10

aqueous solutions. Of the proteins tested, cytochrome c from a bacterium that lives in marine hot springs in Iceland, *Rhodothermus marinus*, was the only protein to catalyse this reaction in a stereoselective way (Kan et al. 2016). By subjecting *R. marinus* cytochrome c to the molecular biology process of "directed evolution', a mutated form of the protein was produced that had much higher activity and strong chemical specificity for forming carbon-silicon bonds. Importantly, these properties were maintained when the *R. marinus* protein was expressed in *E. coli* cells (Kan et al. 2016). The main goal of this research was to investigate if a biological production process could offer a possible substitute for the current chemical syntheses of organosilicon, which are complex, require a high input of energy and often depend on rare or expensive metals. The research has also evoked further discussion on whether conditions for natural organosilicon chemistry might exist somewhere within the Universe.

Moving to broader considerations of where else life could potentially have evolved, the possibility of life outside planet Earth is thought to be restricted to planets with similar climates. For example, it has long been thought that the planet Venus could not harbour life due to its high surface temperatures and the sulphuric acid-rich (90%) composition of its clouds. Although in 2020, certain radio emissions from Venus' atmosphere were interpreted to mark the presence of the gas phosphine (PH_3; considered on Earth a molecular marker for life), this remains highly controversial (Greaves et al. 2020; Voosen 2020).

From knowledge of star systems in our Milky Way galaxy, efforts have been made to calculate how many planets have Earth-like conditions on which potentially advanced forms of life might have evolved. The first calculations were by the astronomer Frank Drake in the early 1960s, leading to the so-called Drake equation. A new approach, published in 2016, set up a different question: "How likely is it that humans are the only advanced, technological life form to have ever evolved?" The calculations yielded a strikingly small probability for this possibility: one in ten billion trillion. On this basis, the authors proposed that it is most likely that other intelligent life forms *have* evolved at some time and place within the Milky Way. Even if alien life forms exist at the present time, it is most likely that they are so far away in space that they are beyond the reach of human communications (Frank and Sullivan 2016). A new analysis from NASA, based on data on exoplanets of the stars observed by the Kepler Space Telescope and the European Space Agency's Gaia Observatory, has estimated conservatively that 300 million rocky planets within the Milky Way have the conditions to support surface water and could potentially be suitable for life. In the context of this estimate, a planet with living organisms could be within 20–30 light years from Earth (Bryson et al. 2020).

Another recent study used Bayesian analysis in combination with data from Earth's fossil record to assign probabilities for the emergence of life and the emergence of human life. The calculations present abiogenesis (i.e., the origin of life from non-living chemicals) on Earth as a fast process, and the evolution of intelligent life as a rare process (Kipping 2020). Another approach has been to use astrophysical information to calculate a possible number of intelligent life forms within our Galaxy, referred to as communicating extra-terrestrial intelligences (CETI). The authors made assumptions on which planets could be considered habitable, based on the known conditions and evolution of life on Earth. They also raised a further assumption that intelligent life should develop within 4.5–5.5 billion years after the formation of a planet; this assumption was again based on what we know of the history of life on Earth. From these proposals, it was calculated that there should be at least 36 intelligent life forms within our Galaxy. In this scenario, and assuming the inhabited worlds to be spread uniformly through the Galaxy, the nearest other intelligent life would be 17,000 light years away: far beyond any possibility of communication (Westby and Conselice 2020).

Within our Solar System, the neighbouring planet Mars is considered the most similar to the early Earth and is the most investigated other planet. A lot is known about the chemistry of its atmosphere and core. The atmosphere of Mars is very thin and consists mainly of carbon dioxide (about 96%) with a little nitrogen and argon, very little oxygen, traces of water and small amounts of methane. Shielding of the surface of Mars from cosmic gamma- and UV-radiation is very limited due to the thin atmosphere. The destructive power of cosmic radiation is seen as a major problem for the possible existence of life on Mars (see also Sect. 3.1). Also, the thin atmosphere leads to a low atmospheric pressure (6–7 mbar), far lower than atmospheric pressure on Earth (1,013 mbar). This is not conducive to life as we know it, because water would not be maintained in the liquid state on the surface of Mars. In addition, Mars is further away from the Sun than the Earth (142 million miles compared to 93 million miles, respectively) and its surface temperatures are much lower (see NASA's Mars website (mars.nasa.gov) for weather reports from its InSight landing craft. For example, on September 1, 2019 the temperature range was a chilly −79C to −28C). There is evidence that liquid water was present on the surface of Mars long ago, but, under the present conditions, surface water appears to be locked up in icecaps and not readily available to support life. However, the indication that water was present on Mars in the deep past has led to much debate on whether life may have existed there previously and whether simple life forms might be present now. For example, it has been suggested that water below the soil surface might undergo freeze/thaw cycles or be in liquid form. This would provide localised environments that could possibly support microorganisms. Soil samples from Mars were found to have a pH of around 7.7, which is very suitable for bacterial growth on Earth, although the different chemical composition of Martian soil must also be considered (Hecht et al. 2009). It is hoped that NASA's Perseverance rover, which landed successfully on Mars in February 2021, will collect samples of rocks and surface sediments that can be analysed for signs of previous or current microbial life.

The detection of methane on Mars (Webster et al. 2015; www.nasa.gov/feature/jpl/curiosity-detects-unusually-high-methane-levels) has raised much debate on the possible sources of this simple organic compound and also on the accuracy of the measurements. Was the methane formed from abiotic sources such as volcanic activity, or could it be a product of past or present life forms on Mars? Because methanogenic Archaea on Earth grow in hypoxic conditions and use carbon dioxide as a carbon source for methane production (Deppenmeier 2002), there is great interest in the idea that methanogenic micro-organisms might be present on Mars. As mentioned, geological and environmental explorations of Mars are in progress with NASA's Mars rovers and other landing craft. A European Space Agency rover that will drill down below the surface in search of signs of life is scheduled for launch in 2022. In parallel, laboratory experiments on Earth have begun to examine if earthly methanogenic micro-organisms can survive under conditions similar to those on Mars. Methogenic Archaea cultured in aqueous conditions under low pressures of 50 mbar or 400 mbar were found to survive and produce methane, and could survive for (at least) a short time under 6 mbar pressure in an aqueous environment (Mickol and Kral 2017). Other experiments have shown that some species remain viable after long periods of desiccation. These and other studies suggest that it might be possible for methanogenic micro-organisms to live below the surface soil of Mars (Cockell et al. 2016).

Many bacteria form dormant structures known as spores that are resistant over long periods of time to harsh environmental conditions and which reactivate later in suitable habitats. It is a point of discussion whether spores could travel from Earth to other planets in rocks or space dust (Nicholson 2009). Another interesting question is whether spores from Earth, that could potentially travel on the surfaces of spacecraft (so-called spacecraft-associated facility surfaces), could survive space travel or Martian conditions. This question has been addressed under laboratory conditions. *Bacillis subtilis* spores were exposed to simulated Martian conditions for 24 h, either without or with Mars-like UV irradiation for 8 h. Even after UV irradiation, some spores survived, through mechanisms involving DNA protection and DNA damage repair. Without UV irradiation, survival was higher, indicating that shielding by space travel equipment, or even shielded niches on Mars, could provide environments potentially habitable by *Bacillis* (Cortesão et al. 2019). Similar experiments have been carried out on fungal spores, lichens and bryophytes and these also indicate a capacity to survive in the short-term (reviewed by Cockell et al. 2016). In addition to establishing that micro-organisms from Earth can survive in the short-term under Mars-like conditions, these experiments provide a sobering indication of the care that is needed to avoid introducing spores or micro-organisms from Earth onto the Moon, Mars or any other destinations during exploratory missions.

Other areas of research focus on identifying and studying bacteria and Archaea that inhabit extreme environments on Earth, such as the polar permafrost or the Atacama Desert in Chile. These are viewed as the environments that are most similar to the surface of Mars. Understanding how and where micro-organisms survive in these environments could help direct any searches for life on Mars to the most suitable candidate regions (Siela and Smith 2019). Given the variability of freeze/thaw cycles measured in permafrost regions, one study has examined the effects of lengthy exposure to freeze/thaw cycles on methanogenic Archaea. Even after 3.5 years of freeze/thaw cycling between temperatures in the range −80C and +55C for various numbers of days, the cells were still capable to begin methane production when placed back into standard conditions (Mickol et al. 2018). Thus, it seems possible that permafrost-like environments on Mars might support life.

In recent decades, additional planet-like bodies have been found in our Solar System. Analysis of the orbits of bodies discovered beyond Pluto, in the Kuiper Belt, has suggested the presence of a large "Planet X" or "Planet 9" far beyond Pluto at the edge of our Solar System (Trujillo and Sheppard 2014; Batygin and Brown 2016); the existence of this planet is hypothetical at present. Modern astronomical tools have also detected, and continue to detect, very many new planets orbiting other stars, i.e., outside the Solar System. These are referred to as exo-planets. Some of these planets are thousands of light years away and have very extreme environments, but others have more moderate environments. Recently, NASA's Transiting Exoplanet Survey Satellite (TESS), that hunts for planets, identified three new exo-planets orbiting a star (designated TOI270) 73 light years away. These planets are in the same size range as Earth and appear interesting for further study (https://exoplanets.nasa.gov/news/1593/tess-scores-hat-trick-with-3-new-worlds/). Another exo-planet, K2-18b, that is 124 light-years away from Earth in the constellation of Leo, was recently identified to have water vapour in its atmosphere and a temperature of around 50 F/10 C. These properties could be compatible with a habitable planet, however K2-18b is much larger than Earth and is expected to have an unfavourable, hydrogen-rich, high pressure atmosphere (Tsiaras et al. 2019). Overall, over 4000 exo-planets have been identified to date, reigniting the question whether other life forms exist within the Universe.

In addition to these efforts, NASA and other bodies are considering how to search for "techno-signatures". i.e., evidence of life elsewhere in the Universe with the intelligence to create technology. Whereas the long-running project SETI@home (SETI stands for Search for Extra-Terrestrial Intelligence) focused on narrow bandwidth radio signals, "techno-signatures" could include radio or laser emissions, evidence for massive structures, or evidence of disturbance of the natural atmosphere. At present, there is no evidence on this point and so researchers can only continue to estimate probabilities for the existence of life elsewhere.

10.2   Possibilities for Emigration from Earth to Other Planets

Another major group of questions in astronomy relate to the possibility of whether humans could travel to and/or live on other planets. The Outer Space Treaty has been in place since 1967 and is presently ratified by 109 countries. This treaty sets out that space is free for exploration and use by all nations, but that no nation may claim sovereignty of outer space or any celestial body, and that any use must be for peaceful purposes. In 2020, NASA announced another legal framework, the Artemis Accords, a set of "Principles for cooperation in the civil exploration and use of the Moon, Mars, comets, and asteroids for peaceful purposes" "to create a safe and transparent environment which facilitates exploration, science, and commercial activities for all of humanity to enjoy" (https://www.nasa.gov/specials/artemis-accords/index.html). Seven other nations signed with the USA as founding members. The framework covers agreements for setting up zones for further exploration and possible mining on the Moon. In October 2020, NASA reported the discovery of water on the Moon (https://www.nasa.gov/press-release/nasa-s-sofia-discovers-water-on-sun lit-surface-of-moon/). Better knowledge of the Moon's resources could help further exploration of the Moon by astronauts, or may lead to new sources of rocket fuel for space missions. The knowledge and technology that would be developed could benefit future efforts to send astronauts to Mars.

Given the great distances involved in any space travel (e.g., travel to Mars takes about seven to nine months), there are major difficulties for humans with the journey itself: for example, the need to launch a rocket that can carry all the necessary supplies and has sufficient shielding against cosmic radiation. Humans on such a rocket will undergo major physiological consequences from an extended period of weightlessness. Most prominently, they will experience loss of bone mass and the atrophy of skeletal muscles. Other known effects include poor blood circulation, damage to the heart and impaired eyesight. The world record holder for the longest single period in space is Valeri Polyakov, who spent nearly 438 days on the Russian Mir space station in the 1990s. The NASA astronaut Peggy Whitson has accumulated a total of 655 days in space (https://www.nasa.gov/feature/nasa-station-astronaut-rec ord-holders/). In 2015, the American astronaut Scott Kelly was monitored during a year on the International Space Station while his identical twin brother, Mark Kelly, provided a control person on earth. Many physiological and cellular parameters were altered during Scott Kelly's time in space. Changes in his expression of some genes, chromosomal inversions, altered numbers of short telomeres and reduced cognitive function continued for at least six months after he returned to earth (Garrett-Bakelman et al. 2019). Only time will tell if there are any other long-term consequences, but it is clear from current knowledge that effects of long-term weightlessness and possible radiation exposure pose major problems to sending humans off on extensive space travel.

One response to these physiological issues has been to raise ideas on how human life could be adapted to make space travel more feasible. A prominent institution that is considering improvement of life, artificial intelligence, and possible settlement of other stars is the Future of Life Institute, Boston (2019). This institute was founded by well-known scientists mainly from astronomy, physics and computer science and includes Max Tegmark. Sir Martin Rees is a member of the advisory board and the late Professor Stephen Hawking was a prominent supporter. Very few biologists participate. Financially, the institute is supported by the highly successful investor Elon Musk, who is well known for introducing the Paypal money transfer system, the production of electric Tesla cars, and as an owner of the space-shuttle producer, SpaceX. This company employs about 2000 people, with the aim of developing low-cost rockets for travel to the universe. The home page of the Future of Life Institute also includes many interesting contributions about the current problems in climate change. The general aim of the institute is to challenge these problems by novel technological advances, for example by artificial intelligence or emigration to other planets. Efforts of the Future of Life Institute include the building and launching of space rockets, or the possible development of out-stations on the Moon or Mars. These are all high technology processes that consume a lot of energy and will require large industrial investments to be developed. Nevertheless, settlement on Mars was offered by the Dutch non-profit organization Mars 1 as a one-way process. This company went bankrupt, but had planned special bunkers to allow humans to live on Mars (Fig. 10.1).

In relation to the problems of space travel, speculative ideas have developed amongst astronomers and physicists, best represented by the publications of Sir Martin Rees (Rees 2017, 2018). These discuss how inorganic machines might be developed as followers of human beings. These entities would be put together by artificial intelligence and would consist of inorganic materials to circumvent the soft and susceptible organic nature of the brain and other parts of the human body. An inorganic brain could also make it possible for these mechanical beings to be connected directly to computer networks. Through these radical modifications it might become possible develop "concious" entities suitable for space travel and to solve the problem of physiological deterioration during space travel. Inorganic entities would also be less susceptible to destructive cosmic gamma-radiation on planets such Mars. A criticism of these ideas comes from biology. We described in Chap. 3 that inorganic chemistry is unable to perform the complex processes of life. Also, most astro-biologists and chemists consider that extra-terrestrial life, if it exists, is likely to depend on the presence of organic (carbon-based) compounds.

NASA is working on a mission concept plan to put humans on Mars by 2033, the Deep Space Gateway and Transport Plan. Another possibility to travel to Mars will

Fig. 10.1 Settlement on Mars according to advertisements of the Mars 1 Company

be offered by the SpaceX Company of Elon Musk (https://www.spacex.com). Flights of Musk's prototype Starhopper have been reported in the media. The company is now working on a large rocket, Starship, which they aim will carry up to one hundred people on flights to the Moon and Mars. In contrast to other rockets, it will be designed for repeated use, suggesting that one goal is a large commercial enterprise for rich customers or corporations. It seems that one way or another humans may arrive on Mars during the twenty-first century. A possible positive effect of transferring humans to other planets might be the chance to preserve the human genome from any global disasters on our own planet, as suggested by Sir Martin Rees (2017).

References

Batygin K, Brown ME (2016) Evidence for a distant giant planet in the solar system. Astron J 151:22. https://doi.org/10.3847/0004-6256/151/2/22

Bryson S, Kunimoto M, Kopparapu R et al (2021) The occurrence of rocky habitable zone planets around solar-like stars from Kepler data. Astronomical Journal 161, 36. https://doi.org/10.3847/1538-3881/abc418

Cockell CS, Bush T, Bryce C, Direito S, Fox-Powell M, Harrison JP, Lammer H, Landenmark H, Martin-Torres J, Nicholson N et al (2016) Habitability: a review. Astrobiology 16:89–117

Cortesão M, Fuchs FM, Commichau FM, Eichenberger P, Schuerger AC, Nicholson WL, Setlow P, Moeller R (2019) Bacillus subtilis spore resistance to simulated Mars surface conditions. Front Microbiol 10:1–16

Deppenmeier U (2002) The unique biochemistry of methanogenesis. Prog Nucleic Acid Res Mol Biol 71:223–283

Frank A, Sullivan WT (2016) A new empirical constraint on the prevalence of technological species in the universe. Astrobiology 16. https://doi.org/10.1089/ast.2015.1418

Future of Life Institute (2019) Institute@futureoflifeinstitute

Garrett-Bakelman FE, Darshi M, Green SJ, Gur RC, Lin L et al (2019) The NASA Twins Study: a multidimensional analysis of a year-long human spaceflight. Science 364:eaau8650. https://doi.org/10.1126/science.aau8650

Greaves JS, Richards AMS, Bains W et al (2020) Phosphine gas in the cloud decks of Venus. Nat Astron. https://doi.org/10.1038/s41550-020-1174-4

Hecht MH, Kounaves SP, Quinn RC, West SJ, Young SM, Ming DW, Catling DC, Clark BC, Boynton WV, Hoffman J, Deflores LP, Gospodinova K, Kapit J, Smith PH (2009) Detection of perchlorate and the soluble chemistry of martian soil at the Phoenix lander site. Science 325:64–67. https://doi.org/10.1126/science.1172466

Jacob DT (2016) There is no silicon-based life in the solar system. Silicon 8:175–176. https://doi.org/10.1007/s12633-014-9270-7

Kan SBJ, Lewis RD, Chen K, Arnold FH et al (2016) Directed evolution of cytochrome c for carbon–silicon bond formation: bringing silicon to life. Science 354:1048–1051. https://doi.org/10.1126/science.aah6219

Kipping D (2020) An objective Bayesian analysis of life's early start and our late arrival. PNAS 117:11995–12003. https://doi.org/10.1073/pnas.1921655117

Mickol RL, Kral TA (2017) Low pressure tolerance by methanogens in an aqueous environment: implications for subsurface life on Mars. Orig Life Evol Biosph 47:511–532. https://doi.org/10.1007/s11084-016-9519-9

Mickol RL, Laird SK, Kral TA (2018) Non-psychrophilic methanogens capable of growth following long-term extreme temperature changes, with application to Mars. Microorganisms 6(2):pii: E34. https://doi.org/10.3390/microorganisms6020034

Nicholson WL (2009) Ancient micronauts: interplanetary transport of microbes by cosmic impacts. Trends Microbiol 17:243–250. https://doi.org/10.1016/j.tim.2009.03.004

Rees M (2017) Unsere Nachfahren werden Maschinen sein, Neue Zürcher Zeitung, October 2017

Rees M (2018) On the future: prospects for humanity. Princeton University Press

Schröder HC, Wang X, Tremel W, Ushijima H, Müller WE (2008) Biofabrication of biosilica-glass by living organisms. Nat Prod Rep 25:455–474. https://doi.org/10.1039/b612515h

Siela AC, Smith SA (2019) Habitability of Mars: how welcoming are the surface and subsurface to life on the red planet? Geosciences 9:0361. https://doi.org/10.3390/geosciences9090361

Trujillo CA, Sheppard SS (2014) A Sedna-like body with a perihelion of 80 astronomical units. Nature 507:471–474

Tsiaras A, Waldmann IP, Tinetti G, Tennyson J, Yurchenko SN (2019) Water vapour in the atmosphere of the habitable-zone eight-Earth-mass planet K2-18 b. Nat Astron. https://doi.org/10.1038/s41550-019-0878-9

Voosen P (2020) Potential signs of life on Venus are fading fast. Science 370:1021

Webster CR, Mahaffy PR, Atreya SK, Flesch GJ, Mischna MA et al (2015) Mars atmosphere: Mars methane detection and variability at Gale crater. Science 347:415–417

Westby T, Conselice CJ (2020) The astrobiological copernican weak and strong limits for intelligent life. Astrophys J 896:58. https://doi.org/10.3847/1538-4357/ab8225

Chapter 11
Outlook

The topics we have discussed in this book focus attention on the increasing risks that threaten the future of life. In summary, natural risks through asteroid hits, severe weather, volcanic eruptions and storms have existed throughout the history of life on Earth and have shaped the course of biological evolution. Human activities and inventions have brought enormous additional risks. Chemical and nuclear weapons and the possibility of nuclear warfare with modern weapons pose huge risks to all forms of life. Humans have modified the natural environment gradually over thousands of years; yet, in the last ~250 years since the first Industrial Revolution, our vast consumption of fossil fuels has led to global warming and massive, rapid changes in the environment of the entire planet. These changes now threaten individual species, the functioning of inter-connected ecosystems, global biodiversity and humans themselves. The risks caused by human activities are also compounded by the huge increase in the human population over the last 150 years. This exacerbates tensions between human groups and nations and puts more pressure from humans onto the rest of the natural world.

The World Economic Forum publishes each year a "Global Risks" report that makes an integrated assessment of all the major risks facing the world in the immediate future. In the 2020 report, climate change and other environmental risks dominated the predicted top ten risks of highest impact for the next ten years (i.e., five out of the ten top risks are environmental and the societal risk of water crises is also strongly related to global warming) (Fig. 11.1). In addition to the environmental and societal factors we have discussed, technological risks from breakdown of information infrastructure and cyber attacks were considered amongst the top impact risks. A category was also included for "climate action failure", i.e., the failure of humans to act quickly enough to reduce greenhouse gas emissions (Fig. 11.1). In contrast, the 2021 report placed infectious diseases as the top long-term risk for impact, with "climate action failure" as the second-placed risk (http://www3.weforum.org/docs/WEF_The_Global_Risks_Report_2021.pdf).

J. C. Adams and J. Engel, *Life and Its Future*,
https://doi.org/10.1007/978-3-030-59075-8_11

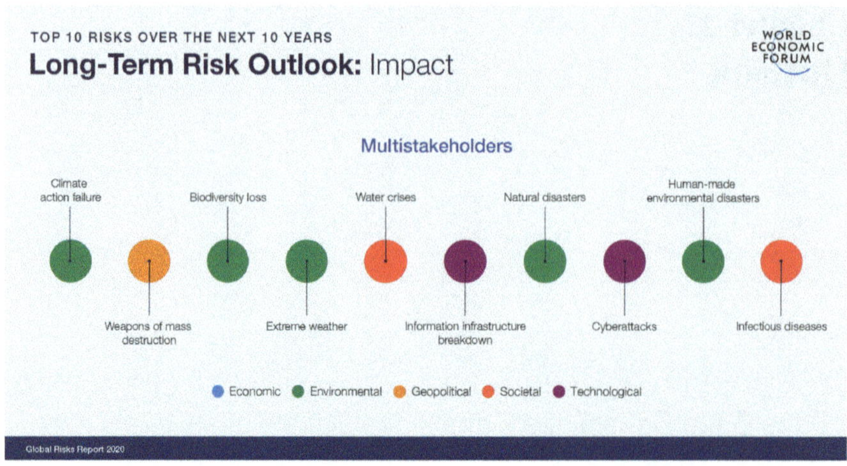

Fig. 11.1 Assessment of the top ten global risks over the next ten years. World Economic Forum, Global Risks report infographic, January 2020

Looking ahead to how we may deal with all these dangerous situations in the future, it is clear that all current measures against use of nuclear, chemical or biological weapons need to be maintained and strengthened. To date, these risks have been controlled by treaties yet nuclear weapons treaties are becoming weaker. The development of weapons of unbelievable efficiency in the military sense is progressing fast and associated risks are increasing exponentially. In addition, illegal groups such as terrorists may in future use weapons of mass destruction. We must also work towards complete bans in international law against weapons of mass destruction and their production, and for the decommissioning of current weapons. The belief that deterrence by nuclear weapons can prevent nuclear warfare in combination with international treaties may in part be true. Going forward, in a less-stable global political climate, this is too uncertain a political method to prevent a global catastrophe.

What can be done to combat the continuing global warming to avoid catastrophic climate changes? The many environmental action protests that took place around the world in 2019 have raised awareness of the need to accelerate the pace and scope of counter-actions. To date, politics and international agreements lag behind the scientific findings. Many political leaders favour a continuation of efficient commercialism and "business as usual", apparently unaware that biodiversity and a healthy environment are vital to human life. The covid-19 pandemic unfortunately shows us how different disasters can interact in unwanted ways; for example, that particulate pollution in the air may have facilitated the spread of the coronavirus. "Green Recovery" or "Green New Deals" are under discussion in various countries and the economic effects of the pandemic may hopefully speed efforts to transition to revised, less polluting, low-carbon economic models. By Novemeber 2020, 126 countries have set themselves, or are considering to adopt, goals to reduce to net-zero emissions These countries account for 51% of the global greenhouse gas emissions (UN Emissions Gap Report 2020).

The Paris agreement of 2015 and subsequently published climate change modeling reports have emphasized that keeping global warming to an average of + 1.5 °C by 2100 is much preferable than an average increase of + 2 °C. Unfortunately, it is likely that the world is already heading for more than another 1.5 °C of warming. Implementing the current pledges of nations to reduce greenhouse gas emissions is calculated to result in a 66% probability of + 3.2 °C of global warming by 2100. To hold emissions to the + 1.5 °C target, countries need to reduce their greenhouse gas emissions by *an average of 7.6% per year*, starting immediately (UN Emissions Gap Report 2020). The importance of this extremely urgent and challenging goal cannot be over-emphasized. Political actions are essential to bring about these deep changes at national and international levels, but we must also act as individuals by taking practical steps to strongly decrease our own greenhouse gas emissions and "carbon-deficit" from home life, travel and work, starting now.

To oppose the negative trends and increasing general risks for life, a change of attitude by human society is needed. The origin of life on Earth is only partially understood in spite of the great development of the biological sciences. Humans have caused many species to become extinct, and so the unique information and biological roles of those species have been lost already. Artificial life has not been achieved and its creation may not prove helpful to solve current problems, as some people believe. True wilderness is shrinking every year and some of the nature we love can now only be seen in nature reservations. We need to think and plan not only for our own and immediate family's benefit but also for the benefit of future generations and for the biosphere in which we all live. Planning for societal resilience needs to become as central as the economic bottom line. The philosophy of environmental ethics underpins the concept of ecocide and climate litigation cases, which argue that public natural resources are not being maintained properly for the benefit of future generations. In changing human attitudes, scientists have a major part to play, not only in collecting data or building models, but also by explaining why it has become so urgent for humans to move away from fossil fuels and to reduce overall consumption (Ripple et al. 2020). For these warnings to be effective, scientists must maintain public confidence in the validity of the scientific method and scientific knowledge. We very much hope that intelligent problem-solving and caring will dominate and ways will be found to reduce the risks of the threatening catastrophes. Prevention of catastrophes must become a dominating aim of nations and individuals.

References

Ripple WR, Wolf C, Newsome TM, Barnard P, Moomaw WR (2020) World scientists' warning of a climate emergency. Biosci 70:8–12. https://doi.org/10.1093/biosci/biz088

UN Emissions Gap Report 2020 (2020) https://www.unep.org/emissions-gap-report-2020

World Economic Forum (2020) The global risks report 2020. https://www.weforum.org/reports/the-global-risks-report-2020

Index

GPSR Compliance
The European Union's (EU) General Product Safety Regulation (GPSR) is a set
of rules that requires consumer products to be safe and our obligations to
ensure this.

If you have any concerns about our products, you can contact us on

ProductSafety@springernature.com

In case Publisher is established outside the EU, the EU authorized
representative is:

Springer Nature Customer Service Center GmbH
Europaplatz 3
69115 Heidelberg, Germany